Where Industry Failed

Water-Powered Mills *at* Harpers Ferry West Virginia

by Dave Gilbert

PICTORIAL HISTORIES PUBLISHING COMPANY
CHARLESTON, WEST VIRGINIA

LIBRARY OF CONGRESS
CATALOG NUMBER 84-60831

ISBN 0-933126-49-2

First Printing June 1984

Typography: Arrow Graphics
Layout: Stan Cohen

Photograph on the front cover shows the remains of the outlet arch of the old cotton factory on Virginius Island. The back cover shows the remains of additional arches on the island.
Courtesy Deborah Piscitelli

Cover Artwork: Mary Beth Percival

PICTORIAL HISTORIES PUBLISHING COMPANY
4103 Virginia Avenue S.E.
Charleston, West Virginia 25304

CONTENTS

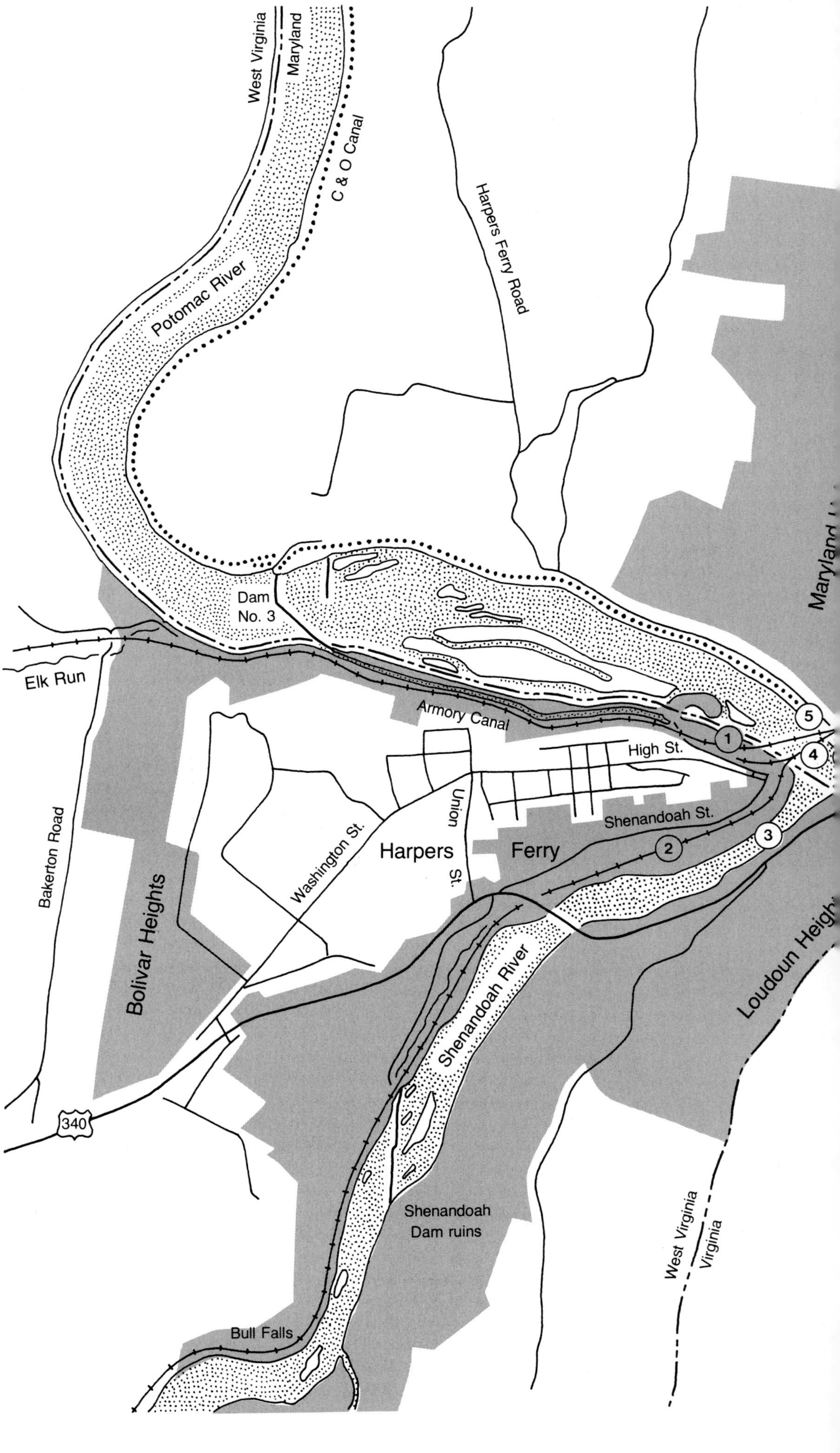
West Virginia
Maryland
C & O Canal
Harpers Ferry Road
Potomac River
Dam No. 3
Maryland Heights
Elk Run
Armory Canal
High St.
Union St.
Shenandoah St.
Washington St.
Harpers Ferry
Bakerton Road
Bolivar Heights
Loudoun Heights
Shenandoah River
340
Shenandoah Dam ruins
West Virginia
Virginia
Bull Falls
1
2
3
4
5

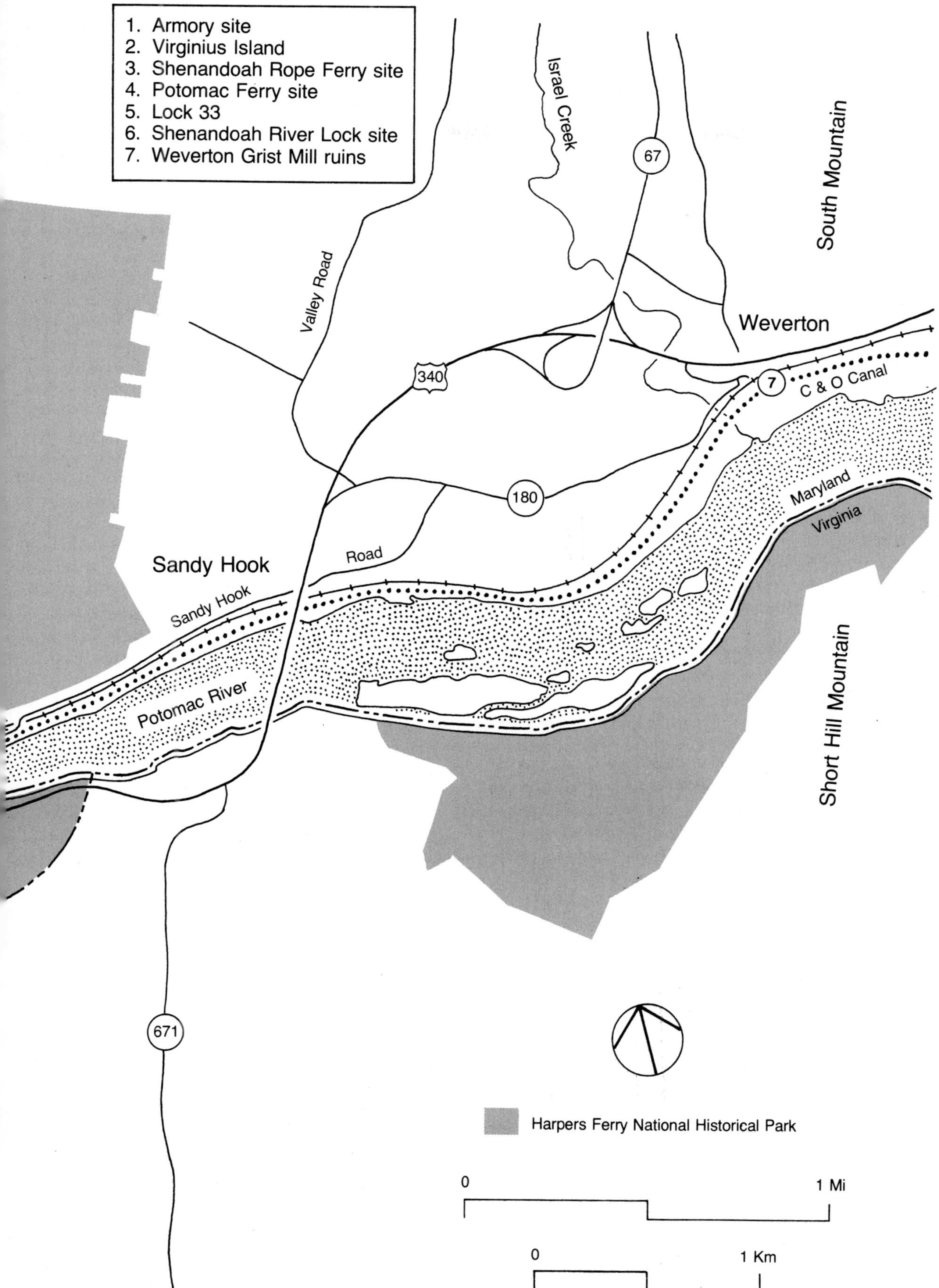

Harpers Ferry Today

ACKNOWLEDGMENTS

Several people made indispensable contributions to this book. I'm particularly indebted to Hilda Staubs of Harpers Ferry National Historical Park for making the park's extensive reference library, research reports, and photo collection available to me.

For information on turbines and other machinery, I owe considerable thanks to William E. Worthington, Jr., of the National Museum of American History in Washington, D.C.; to Florence Brigham of the Fall River Historical Society in Fall River, Mass.; to Gary J. Arnold of The Ohio Historical Society in Columbus, Ohio; and to Albert V. James of the National Capital Region of the National Park Service in Washington, D.C.

For very helpful material on the Shenandoah Pulp Company, I'm grateful to Helen E. Arnold of the Potomac Edison Company in Hagerstown, Md.; to Lewis Nichols of Harpers Ferry; and to W.H. Henchey of Westvaco Corporation in New York City.

I'm also indebted to the Geography and Maps Division of the Library of Congress; to the National Archives; to the United States Patent Office; to the Washington County Library in Hagerstown, Md.; and to the Washington County Historical Society.

Finally, I owe a great deal of thanks to Deborah Piscitelli of the Harpers Ferry Historical Association and to Jim Jenkins of Charles Town, W.Va., for their timely and helpful editorial comments.

Unless otherwise noted, all photographs are by the author.

About the Author

Dave Gilbert is a writer, editor, and professional river guide who came to Harpers Ferry in 1975. After several seasons of rafting and canoeing, Dave's curiosity about old dams and crumbling mill ruins sparked serious research, and led ultimately to this book. Other books by the same author include *A Walker's Guide to Harpers Ferry, The American Bicycle Atlas,* and the *Greater Washington Area Bicycle Atlas.*

Chapter One

INTRODUCTION

American industrial history is foremost a history of human accomplishments. From the earliest Colonial days hard work, ingenuity, and daring investments of capital spawned small community industries and ultimately created an expansive and highly productive factory system. Basic to this development was our nation's rapidly expanding domestic market following the American Revolution. This, combined with improvements in transportation—initially roads and canals—brought people and products ever closer together. In the decades preceding the Civil War, the rapid growth of American railroads stimulated industry even further, accelerating the expansion of old factories and encouraging the construction of new ones.

As demand for finished products was growing by leaps and bounds, and improved transportation was speeding delivery to customers everywhere, manufactories introduced machinery and other improvements to quicken production. This development is reflected in Patent Office records, which from 1790-1811 reported an average of just 77 inventions annually. Compare this number to the year 1830, when the Patent Office issued 544 patents. In the decade 1841-1850, 6,460 patents were recorded, followed in the next decade by a whopping 25,250 patents.

Extension of mechanization and improvements in production methods were cornerstones in effecting a factory system in America. A move toward progressively larger establishments attended these developments, requiring access to additional power to drive the belts and shafts of the new machinery. Water power, which drove mills in Greece and Rome as early as the First Century B.C., was still the primary source of power in the states prior to 1870. To harness this power, sizable waterworks—milldams, headgates, and millraces—were erected, conveying falling water to the water wheels of our first factories. Access to falling water thus determined the location of expanding industries from the start. Along major streams and rivers, substantial manufactories and human settlements quickly sprang up.

As this brief overview suggests, an unfortunate circumstance in a study of industrial history is its frequent generality and ponderous impersonality. To consider industrial history as a true history of human accomplishments, we must narrow the context of our story to a particular time and a particular place. In this manner we can weigh specific human contributions and individual human labors—the building blocks of America's entire factory system. Focusing, for instance, on Harpers Ferry in the 19th Century, we are able to define a specific physical, social, economic, and technological setting, see more clearly the interrelationships of people and events, and give our story a certain life and vigor. Individuals like Robert Harper, George Washington, John H. Hall, Lewis Wernwag, Abraham Herr, and Thomas Savery—some well-known and some not—each assume importance for their particular contributions or personal achievements. Together, they are participants in a story spanning almost 200 years.

The evolution of industry at Harpers Ferry mirrored in many important respects the rise of the factory system in America. The various and interrelated factors that determined early settlement patterns, transportation corridors, and the growth of manufacturing in general can be traced at Harpers Ferry in particular. Here, on a few hundred rocky acres bounded by the Potomac and Shenandoah rivers, were born impressive examples of human success, disappointing instances of human failure, and an enduring picture of the permanence of nature.

From modest beginnings, when just a handful of settlers patronized a gristmill and a sawmill here in the late 1700s, Harpers Ferry grew into a sprawling industrial town. By 1860, an expansive federal armory and arsenal, textile mill, flour mill, sawmill, iron foundry, machine shop, carriage manufactory, and 43 shops and businesses operated at Harpers Ferry. These various establishments served a local population of almost 3,000 people and other growing domestic markets nearby.

The rapid growth of industry at Harpers Ferry between the years 1790-1860 can be attributed to many of the same causes that affected the development of the factory system across America. The partial shutting off of European imports, for instance, during the American Revolution, the Embargo and Nonintercourse acts, and the War of 1812 stimulated domestic manufacturing in general, and the textile, iron, and arms industries in particular. In the Harpers Ferry area, Antietam Iron Works and Keep Triest Furnace expanded to meet growing demands for finished iron products formerly imported from England. Muskets and other firearms, which had previously been procured from European manufactories, were also being made in America. The location of a federal armory at Harpers Ferry was a result of this change.

Harpers Ferry as it appeared in 1896. Already the presence of industry had faded. Of the town's antebellum manufactories, only the old cotton factory building (upper left) remained standing along the Shenandoah River. In the right foreground, the armory grounds had recently been covered with a new railroad embankment. National Park Service

Another stimulus to the development of industry in this country was our abundance of raw materials, particularly cotton, wood, and iron, and of water power. The Harpers Ferry area, which possessed plentiful quantities of wood, iron, and water power in the early 1800s, attracted new settlers and new industry at a substantial rate. The Census Office, in its *Reports on the Water-Power of the United States* issued in 1885, found that the waters of the Potomac and Shenandoah rivers here could produce in peak periods up to 18,350 gross horsepower. The report further claimed that as an industrial site Harpers Ferry was "probably the most favorable one on the river."

On yet another front, skilled and unskilled European laborers, immigrating to America in continually increasing numbers, provided an ample work force for new and expanding factories. Among these many immigrants, German and Irish settlers found their way to Harpers Ferry either by way of Pennsylvania or as indentured laborers performing construction work on the Patowmack Canal and later on the Chesapeake & Ohio Canal.

Government aid through protective tariffs also helped stimulate domestic industries, and there appeared substantial amounts of capital for investment in manufactories. Federal funding of the Harpers Ferry Armory, which provided for the continuing expansion of this arms-making establishment prior to the Civil War, as well provided capital for or encouraged private investment in other local industries—an iron foundry, machine shop, sawmill, and flour mill, for example.

As the factory system in America continued to develop through the period 1790-1860, so, too, industry at Harpers Ferry tried to keep pace. At one time or another, factories here employed the machines of some of this country's most important inventors: the gun stock lathes of Thomas Blanchard, the textile manufacturing improvements of Eli Whitney and Samuel Slater, the flour milling innovations of Oliver Evans, and the iron turbines of Uriah Boyden and James Leffel. John Harris Hall, working for the U.S. Government right here at Harpers Ferry, made his own contribution to the factory system by perfecting interchangeable mechanisms and standardization of parts in arms manufacture. Hall's accomplishments laid the foundation for assembly line production in America.

Just as the development of individual industries

George Washington Grist Mill near Mt. Vernon is typical of many 19th century gristmills that were built along the Potomac River. This particular building was erected at Front Royal, Va., in 1814 and moved to its present location by the Civilian Conservation Corps during the Depression.

at Harpers Ferry provides us with specific examples of a nationwide trend, so, too, improvements in transportation stimulated factories here as elsewhere. From Colonial times small towns and community industries relied on rivers not only for water power, but to transport their goods as well. In time came canals—at first man-made improvements to existing riverbeds, and later artificial waterways with lift locks and towpaths. The Patowmack Canal was an example of this first type of waterway; the C & O Canal was its more reliable successor. Both canal systems during their respective years of operation linked Harpers Ferry with Georgetown and Alexandria—two 19th century seaports in the Potomac estuary. Turnpikes—roads on which a toll was charged by the mile travelled—connected towns not readily accessible by river or canal. Harpers Ferry served as a junction for three such roadways: the Harpers Ferry, Charles Town & Smithfield Turnpike, the Frederick & Harpers Ferry Turnpike, and the Hillsborough & Harpers Ferry Turnpike. When the Baltimore & Ohio Railroad linked Harpers Ferry with Baltimore, yet another stimulus was given to local industry to produce and distribute its goods.

The story of human accomplishments at Harpers Ferry is balanced by compelling tales of human failure. For even as manufactories here were expanding up the shorelines of the Potomac and Shenandoah rivers, steam was replacing water as the primary source of power in America. No longer bound to streams and rivers, factories gradually relocated closer to raw materials—coal, iron, and timber, for instance—or to expanding domestic markets. Those factories that remained riverbound, as did the industries of Harpers Ferry, suffered from increasingly unreliable water supplies and devastating floods—by-products of extensive lumbering in the Potomac and Shenandoah river valleys.

Human conflict also turned the tide against industry at Harpers Ferry. John Brown's raid and the subsequent carnage of the Civil War destroyed property, displaced residents, and sent investors with capital generally looking elsewhere. A few investors persevered, however, and industry made a modest comeback at Harpers Ferry in the years

Most of the old gristmills along the Potomac River have been totally obliterated or left in ruins by the devastations of time and weather. These ruins, identified as "A large stone Flour Mill" on an 1849 survey map, stand at the foot of Short Hill Mountain in Loudoun County, Va.

Today, Harpers Ferry is a National Historical Park, where students can learn about man's accomplishments and mistakes of the past two centuries.

following the war.

There are indeed lessons to be learned from the story of Harpers Ferry. One lesson is the durability of nature. It was the beauty of the place that prompted Thomas Jefferson to write in his *Notes on the State of Virginia* in 1785: "The passage of the Patowmac through the Blue Ridge is perhaps one of the most stupendous scenes in Nature." But just the same the practical advantages of the site drew many more who erected factories, cut timber to fuel furnaces and forges, and dammed the rivers with rock and rubble. So great became the intrusion of industry that in 1835, two British travellers described Harpers Ferry as "a most abominable village," pervaded by the "smell of coal smoke" and the "clanking of hammers" from factories.

Nevertheless, time has smoothed the rough edges. Today sycamore, yellow poplar, box elder, and silver maple have reclaimed the riverbanks. Riparian birdlife, alive with song and chatter, flourishes throughout the Harpers Ferry area. And the rivers themselves have endured with much of their original vigor.

That man's presence here, as elsewhere, is precarious and impermanent is also apparent. From a peak of almost 3,000 residents in 1860, today the combined population of Harpers Ferry and Bolivar is less than half that number. Where factories and houses once stood, only crumbling stone foundations, scattered bricks, rusting metal, and silted millraces remain. Spared from successive waves of habitation and development in the present century, these ruins, much as the buried walls of Pompeii, continue to tell a story of human accomplishment and of human failure.

To understand the lessons of Harpers Ferry, imparted from another context of time and place, is to heed the words of Eric Sloane in *Our Vanishing Landscape:* "When a man has lost sight of his past, he loses his ability to look forward intelligently." Ultimately, the story of Harpers Ferry is a saga of man and nature, relevant to the present, and, we hope, to the future.

So our story begins.

Chapter Two

THE LAND & ITS SETTLERS

In 1747, Robert Harper was engaged by some members of the Society of Friends to erect a meeting house on Opequon Creek near the present site of Winchester, Va. Harper, an accomplished architect and millwright, was a native of Oxford, England, where he had been born in 1703. At the age of 20, Harper emigrated to Philadelphia. Here he prosecuted various building projects, and here he was commissioned by the aforementioned Quakers to build their meeting house in Virginia.

Travelling to the Shenandoah Valley in 1747, Harper was persuaded by a certain trader he met in Maryland to cross the Potomac River not at Pack Horse Ford near the village of Antietam, but to proceed more directly by a ferry at The Hole. Here, where the Potomac and Shenandoah rivers converged, Harper met Peter Stephens, a trapper and trader who had squatted at The Hole for 13 years. Harper liked the place. With a millwright's eye he saw great potential for harnessing water power for mills. As an architect, he found here abundant stone for building material. In addition, he reasoned a fair income could be derived from the operation of the ferry.

For 13 British guineas, Harper purchased from Stephens his log cabin, corn patch, and ferry equipment. Within just a few short years, the place became known as "Shenandoah Falls at Mr. Harper's Ferry." Today we know it simply as Harpers Ferry.

The natural setting of this place was surely as impressive to Robert Harper in 1747 as it is to us today. But while we might indulge in such emotions as awe and beauty, Harper was captivated by more practical observations. In particular, falling water issuing from long rapids on the Potomac and Shenandoah rivers challenged Harper's skills as a builder and millwright. With the proper arrangement of milldams and headraces, he knew he could divert water from one or both rivers and, with the falling water, drive the water wheels of a variety of mills. Such a location, where rapids are created by irregularities in geological formation, was called in Colonial times a mill-seat.

The very pattern of geological formation that created these long rapids resulted from natural processes acting over millions of years. During the Paleozoic Era, which commenced about 600 million years ago, sediments from ancient land masses accumulated in a great trough some 2,000 miles long and 300 miles wide. Called the Appalachian geosyncline, this trough extended northeast to southwest, from Newfoundland to present-day Alabama. The Harpers Ferry area lay along the eastern edge of this trough. A broad, shallow sea covered the Appalachian geosyncline, and here sediments—principally sand, silt, clay, and gravel—accumulated to a depth of 30,000 to 40,000 feet. The weight of so much sediment, which continued to collect and subside over millions of years, slowly compacted and cemented layers of erosional debris into sedimentary rock. Sand particles became sandstone, clay and silt were compacted into shale, and the lithified remnants of ancient sea creatures—trilobites, brachiopods, corals, sponges, and other marine organisms which thrived in shallow water—became limestone.

Beginning about 230 million years ago, though time on such a vast scale is notoriously imprecise, forces in the earth's crust drove the strata of the Appalachian geosyncline westward. Sedimentary layers of sandstone, shale, and limestone were squeezed and compressed into a series of parallel folds and thrust faults. It was this ponderous process, called the Appalachian Revolution, that created the Appalachian Mountains. By the end of the Paleozoic Era, some 225 million years ago, this compression and folding of the rock strata was completed.

In the Harpers Ferry area, crust forces exerted during the Appalachian Revolution compressed the sedimentary rock and subjected this strata to intense heat and pressure. Under these powerful physical forces rocks metamorphosed, their internal structure being altered or altogether changed. Sandstone was transformed into quartzite (known locally as *Weverton sandstone* or *Weverton quartzite*) and shale became a very hard schist or phyllite (alternately called *Harpers shale, Harpers phyllite,* or *Harpers formation*).

As the rock strata here was compressed and immense heat and pressure was applied, a massive fold in the earth's crust was raised and overturned westward. Geologists today call this huge fold the South Mountain anticlinorium. This grand formation, which extends north-south from Pennsylvania into Virginia, is comprised of steeply overturned and recumbent folds. Second order folds within the anticlinorium include South Mountain, Short Hill Mountain, Elk Ridge, Loudoun Heights, and Bolivar Heights. The folded strata of *Harpers shale* and *Weverton quartzite* uniformly rise

Folded ledges of Harpers shale *cut across the Potomac River just above Harpers Ferry. These rapids have alternately been called House's Falls, Harpers Ferry Falls, and, most recently, The Needles. The old Armory Canal on the far left conveyed water from Dam No. 3 to the shops of the musket factory.*

westward.

The massive, weather-resistant *Weverton quartzite* is the principal element comprising the mountain ridges around Harpers Ferry. This particular formation is from 200-300 feet thick. The rock is characterized by clear to milky quartz grains and is generally whitish in appearance. *Harpers shale* is comprised of a series of shales and sandstones which have been intensely altered and folded, resulting in a texture much like schist, slate, or phyllite. This formation is about 2,000 feet thick and is usually dark blue-gray in color. Lesser outcrops of gneiss, greenstone, sandstone, and dolomite also occur through parts of the Harpers Ferry area.

Following the close of the Paleozoic Era, a long span of erosion commenced through the Triassic, Jurassic, and Cretaceous periods. The result of this process was a flat plain of erosional debris, commonly called the Schooley peneplain. Then, beginning about 70 million years ago, cycles of uplift and erosion occurred. River valleys were now cut between durable mountain ridges. Streambeds entrenched themselves into softer strata. At about this time the Potomac Watergap and its several rapids began to form as the Potomac River slowly cut a path through the Blue Ridge Mountains. Thomas Jefferson pondered this long process in his *Notes on the State of Virginia:*

> *The passage of the Patowmac through the Blue Ridge is perhaps one of the most stupendous scenes in Nature. You stand on a very high point of land. On your right comes up the Shenandoah, having ranged along the foot of the mountain a hundred miles to seek a vent. On your left approaches the Patowmac in quest of a passage also. In the moment of their junction they rush together against the mountain, rend it asunder and pass off to the sea. The first glance of this scene hurries our senses into the opinion that this earth has been created in time. . . ."*

Time indeed. The Potomac Watergap and its surrounding topography took their final shape about 11 million years ago, when the last major uplift occurred.

The pattern of uniformly folded and intensely resistant rock ledges at Harpers Ferry thus created substantial rapids and, for Robert Harper, an attractive mill-seat. From the present site of Dam No. 3 on the Potomac River (see map page iv), and from about Bull Falls on the Shenandoah River, upturned ledges of *Harpers shale* extend downstream to the confluence of the rivers. Along the Potomac River here, the water descends some 25 feet. Along the

Shenandoah River adjacent to Virginius Island, the water descends about 14 feet. Below Harpers Ferry, exposed ledges of *Weverton quartzite* form yet more rapids in the Potomac Watergap.

The terms "head" and "fall" were used by millwrights from Colonial times onward to denote the difference in height between supply water in a mill's headrace and waste water in its tailrace. In particular, "head" was all the water above the point of impact on a water wheel; "fall" was all that water below the point of impact. "Head" and "fall" together, then, indicated the total descent of a river over waterfalls or rapids at a particular mill-seat.

Of equal importance to the combined "head" and "fall" available at a mill-seat was the volume of water and the reliability of its flow. Consequently, Robert Harper and millwrights after him had to look at the important association between each river's hydrologic cycle and the size of its respective drainage basin.

Above Harpers Ferry, the Shenandoah River comprises a drainage area of 3,054 square miles. The Potomac River above Harpers Ferry, including its North and South branches, comprises a drainage area of 6,380 square miles—more than twice that of the Shenandoah. Precipitation today in both drainage basins averages 38 to 40 inches annually.

The Shenandoah Valley is underlain by carbonate rocks, principally limestones and dolomites. The cavernous limestones in particular serve as important sources of groundwater. During prolonged droughts, limestone aquifers may contribute from 50% to as much as 85% of the flow of the Shenandoah River. This natural circumstance, combined with a stable forest cover which existed throughout the Shenandoah Valley in the late 1700s and early 1800s, helped maintain a certain durability in the flow of the river. Nonetheless, due to the relatively small size of the Shenandoah drainage basin and the steep gradients of its tributaries, fluctuations in streamflow were not unusual even in Colonial times. When forests were cleared for agriculture and trees were felled to fuel iron furnaces and blacksmith forges in the 19th century, variations in streamflow were greatly aggravated.

As regards the Potomac River above Harpers Ferry, its sizable drainage area results in a discharge more than twice that of the Shenandoah. Still, its streamflow has always been subject to wide variations. Describing the Potomac River, its North and South branches, and its several tributaries, the Census Office wrote in its *Reports on the Water-Power of the United States* in 1885:

> *Their chief peculiarity lies in the great fluctuations to which their flow is liable. The rain falling on the mountains is shed rapidly into the water-courses by the steep side slopes leading to the narrow valleys below, and there being few lowlands to overflow, and so to store the freshet water, and no lakes whatever in the region, these streams, and with them the Potomac river, are subject to very sudden and heavy freshets, while in dry seasons their discharge becomes very small.*

Indeed, as early as 1748 and 1753, freshets sent Robert Harper scampering for high ground. But despite these occasional freshets and wide seasonal fluctuations in streamflow, millwrights found the discharge rates of the rivers here adequate and the possibilities for developing the water power great. Besides, the disadvantages of water power resulting from this erratic seasonal supply were of minor consequence owing to the same seasonal limitations of inland transportation; during the prerailway era, raw materials and finished goods could only be transported in quantity when water levels were appropriate for navigation by river craft.

After the railroad arrived at Harpers Ferry and commerce became reliable year-round, still water-powered industry endured and even flourished. This was due in part to new water wheel technologies and to the mill-owner's reluctance to do away with a familiar source of power. Just the same, water power worked; the Census Office, in its 1885 report, still claimed that "there seems to be no reason why a large and fine power could not be utilized here. The site is probably the most favorable one on the river."

Streamflow characteristics at Harpers Ferry, then, were originally determined by natural processes acting over millions of years. Settlement patterns, on the other hand, were first affected by a land grant made less than 100 years before Robert Harper arrived. In 1649, Charles II of England, exiled from his throne by Cromwell and the Parliamentary Commissioners, bestowed upon seven companions in exile that part of Virginia "bounded by and within the heads" of the Potomac and Rappahannock rivers. In 1661, after Charles II was restored to his throne, the land grant was enrolled and upheld in all its terms. Finally, in 1688, Thomas Lord Culpeper obtained a renewal of the grant in perpetuity from James II. Culpeper's property was known in the Colonies as the Northern Neck.

As the terms of the land grant stated, the Northern Neck extended from tidewater on the Chesapeake Bay to the Allegheny Plateau. To the north, Culpeper's estate bordered on the Potomac River to the extreme southwestern tip of Maryland. To the south, this property bordered on the Rappahannock River and on an imaginary line drawn from its headwaters to the headwaters of the Potomac.

The Northern Neck was one of many land grants made by European monarchs to either private joint-stock companies or individual proprietors. Because

John Warner's map of the Northern Neck Proprietary. Warner accompanied a survey party charged with determining the exact boundaries of Lord Fairfax's property in 1736-1737. Four editions of Warner's map are known, this being the fourth and final edition, issued sometime after 1747.

Library of Congress

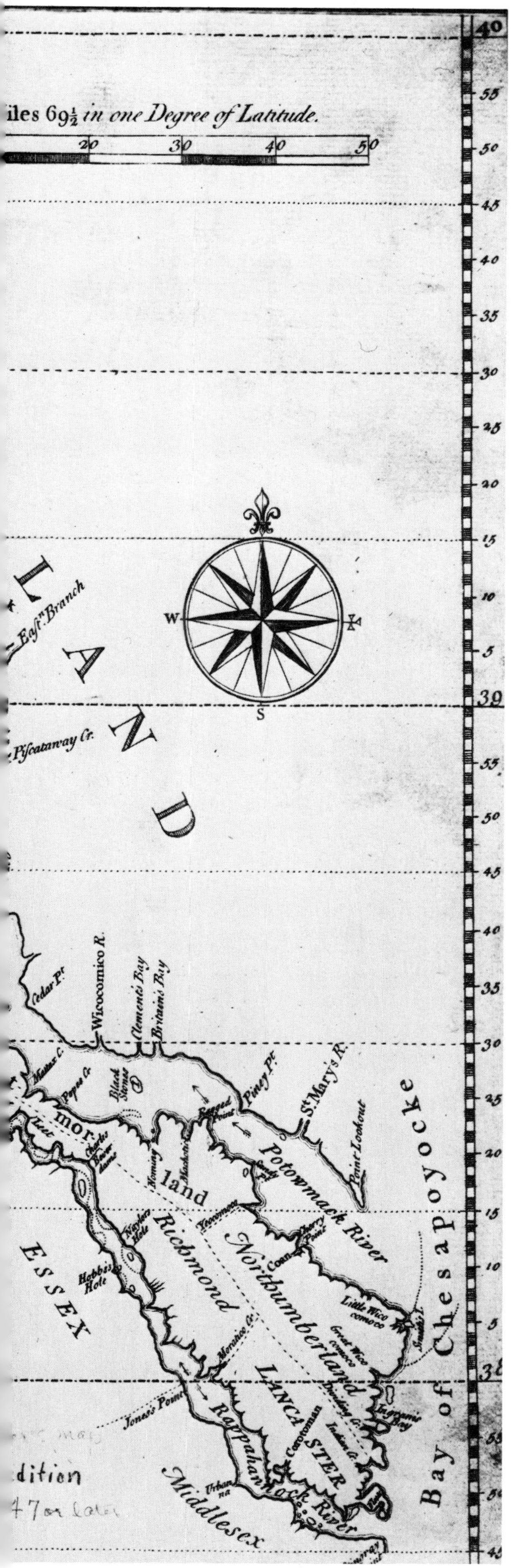

feudalism was the existing land system in Europe, inevitable attempts were made to transfer this system to land grants in the Colonies. When Thomas Lord Fairfax, Culpeper's grandson, came to the Colonies in 1736 as Proprietor of the vast Northern Neck estate, he immediately set about collecting quitrents, or rents, from his tenants. Those unlawful squatters who would not pay he tried to evict. But as Fairfax was to shortly learn, his authority as well as that of the Crown was too weak and too distant to enforce obedience on so large an expanse of property. Indeed, several squatters, among them individuals who endured the hardships, labors, and dangers of frontier life, felt the land was their's simply by right of occupation.

Early records indicate that settlers began to occupy land on the Northern Neck beyond the Blue Ridge Mountains some 30 years before Fairfax came to the Colonies. Wrote Joseph E. Johnson of Philadelphia: "In 1707 traders identified with Pennsylvania made some kind of a settlement near the junction of the Shenandoah with the Potomac. . . ." Just a few miles upstream from here, at Pack Horse Ford, German and Scottish-Irish settlers crossed the Potomac River into Virginia during the 1720s and 1730s. These many immigrants were lured into the Shenandoah Valley from Pennsylvania by the prospect of cheap and abundant land. It is little wonder that they so vigorously opposed Fairfax's demands for rent.

The Germans' first settlement in the Northern Neck, dating from about 1727, was called Mecklenburg and was just a short distance from Pack Horse Ford. In 1734, the land on this site was patented to Thomas Shepherd, and the town's name was eventually changed to Shepherdstown. In 1731, Quakers, who later engaged Robert Harper to erect their meeting house, settled on Opequon Creek. At about the same time the German Hans Jost Heydt, or Joist Hite, settled near the Quakers with 16 families. In 1733, the German trapper and trader Peter Stephens settled at The Hole.

Frustrated in his attempts to deal with certain illegal squatters, among them Peter Stephens, Fairfax returned to England to press his grievances in the courts. His effort brought little result, however, and in 1747 Fairfax came back to Virginia. The following year he took up residence in the Shenandoah Valley at Greenway Court, near Berryville. Here he lived until his death in 1781. Fairfax's quarrels with squatters continued up to the American Revolution. Finally, in 1779, the Northern Neck Proprietary became vested in the Commonwealth of Virginia.

So the stage was set when Robert Harper met Peter Stephens in 1747. Harper, aware of Stephens' precarious tenure over land at The Hole, was not satisfied to only purchase his log cabin, corn patch, and ferry equipment. The following year he travelled to Greenway Court to obtain from Fairfax clear title to the land. In March 1748, Fairfax, possibly prompted by Harper, instructed a surveying party led by James Glenn to lay off tracts in the lower Shenandoah Valley, including the area around the confluence of the Potomac and Shenandoah rivers. George Washington, just 16 years old, and George William Fairfax, Lord Fairfax's cousin, accompanied this surveying party. Finally, on April 25, 1751, Harper obtained a patent to 125 acres.

Robert Harper and his wife moved into Stephens' log cabin, which stood at the present corner of Potomac Street and Shenandoah Street. There is evidence that Harper immediately put his building skills to work by erecting a gristmill. He announced his new establishment with the following notice:

> *To the farmers in Maryland and over the Blue Ridge in Loudoun County. This is to certify that all persons bringing grist to my mill, under the charge of William Griffith, will be ferried over the Potomac and Shenandoah rivers free of expense.*

Harper as well operated the ferry. On July 27, 1757, to insure the continued and proper operation of ferry service across the Potomac, Harper purchased a 20-acre tract of land on the Maryland side of the river known as "Dear Bought." Harper later profited from an act of the Virginia General Assembly, which in 1761 granted him the right to establish and maintain a ferry across the Potomac River. Of course, Harper had been operating a ferry here since 1747, and Peter Stephens before him offered the same service. But by the Virginia act a concession had been awarded, giving Harper the exclusive right to ferry foot and wagon traffic across the Potomac. Harper and his heirs would own these lucrative ferry rights for almost 80 years.

In about 1775, Harper built a new house a half-mile up the Shenandoah River. Then, shortly after moving in, he began to erect a large stone house on the side of a hill above what is now called High Street. Owing to a scarcity of skilled labor during the American Revolution, however, work on this new house proceeded slowly. Finally, in 1782, after eight years of construction, this house was completed. But in October of that year Harper died, and he never occupied the house. Today, the Harper House is the oldest surviving building in Harpers Ferry.

Robert Harper was buried just above town on land set aside in his will. His tombstone, worn by the weather of more than two centuries, is illegible today. But the legacy he left is comprised of more than just a house and a town's name. It was this English-born builder who, by exercising his talents as a practical businessman and skillful millwright, first exploited the site for its economic worth and industrial potential. These accomplishments may seem crude today when placed against the spectacular setting. But be reminded that in Robert Harper's time there were few settlements and almost no industries beyond the Blue Ridge Mountains. Harper truly carved a foothold out of a wilderness and endured through the simple virtues of hard work and enterprise. By the time of his death, the tone had been set for more than a century of industrial and commercial development. Harpers Ferry was indeed primed to share in the growth of the nation at large.

The Harper House, erected by Robert Harper between 1775-1782, is the oldest surviving building in Harpers Ferry. For about two decades after Harper's death, the house was leased as "The Tavern," and hosted such notable guests as Thomas Jefferson and George Washington.

Chapter Three

EARLY INDUSTRY AROUND HARPERS FERRY

Between the years 1734-1739, Thomas Shepherd erected a two-story stone gristmill in the present community of Shepherdstown, W.Va. Called, appropriately, Shepherd's Mill, this building was most probably the first of its kind in the Shenandoah Valley to grind wheat into flour and corn into meal using water power.

The operation of a gristmill from Colonial times on combined mechanical ingenuity with a miller's keen sense for producing a fine, hearty flour. Wooden gears transmitted power from a vertical water wheel to horizontal burr-stones (see diagram, page 16). As well, the arrangement of these gears quickened the speed of the burr-stones, which generally rotated five times as fast as the slower water wheel. The axle, or main shaft, of the water wheel was attached to a parallel face wheel housed in the basement of the mill. Wooden cogs fitted on this face wheel engaged a much smaller gear mounted at the base of the vertical millstone spindle. This gear was usually a lantern pinion—a ring of staves mounted between two hardwood discs. The millstone spindle connected the drive of the water wheel to the burr-stones on the floor above.

Millwrights chose their wood with care. Both wheels and shafts were customarily constructed of oak. For added strength and durability, wheels were laminated using two layers of oak arranged so their grains crossed at right angles. Cogs were shaped from hardwoods such as apple, hornbeam, or beech. An application of tallow—rendered animal fat—smoothed the meshing of the gears.

The burr-stones were the heart of the gristmill and the machinery the miller attended to with the most care. An upper runner stone and a lower, fixed bed stone were housed in a round wooden stone vat. The runner stone was mounted on top of the millstone spindle, which protruded through a neck bearing in the center of the bed stone. A strong, curved iron bar, called a balance rynd, was fitted across the center hole, or eye, of the runner stone. In the middle of the balance rynd was a socket called the cock eye. The top, or pivot point, of the millstone spindle was called the cock head. This cock head engaged into the cock eye and supported the full weight of the runner stone atop the millstone spindle. A cast iron driver, which was attached to the millstone spindle just below the cock head, engaged into recesses on either side of the eye of the runner stone. It was this driver that caused the runner stone to turn when the water wheel was in use.

Grain—corn, rye, or whole wheat—fell from a bin in the loft of the mill into a wooden hopper mounted above the burr-stones. From this hopper grain poured down a slanted spout, or shoe, into the eye of the runner stone. The stones, dressed with a radial pattern of flat lands and grooved furrows, sheared the bran shell from the grain like scissors as it moved from center to circumference.

Because too much speed, pressure, or heat could ruin the texture of the flour, a miller kept his ear tuned to the noise of his machinery and periodically checked the flour for coarseness or fineness. Oliver Evans, who wrote *The Young Mill-wright and Miller's Guide* in 1795, claimed that proper grinding did not result from any specific rule but from "a knowledge depending upon that skill which is only to be obtained by practice." The sharpness of the burr-stones, how close together they were set, the speed of the stones, and the rate at which grain was fed to the stones all affected the quality of the flour.

Evans' book, which appeared in 15 editions from 1795 to 1850, was a comprehensive technical manual. In it were instructions for choosing and building water wheels, selecting appropriate gearing arrangements, dressing millstones for the best grinding effect, and otherwise erecting and operating a successful gristmill. Evans as well introduced innovative supplementary machinery to speed and simplify the manufacture of flour (see Chapter 9).

The most popular millstones in early America were French burr-stones, composed of "burrs," or small pieces of quartz from the Marne Valley of northern France. These small burrs were pieced together, joined with cement, and then bound with round iron hoops. A good pair of burr-stones could grind an average of about 50 tons of grain before redressing was required. When that time came a stone dresser, after lifting the runner stone with an iron stone crane and then turning it over, carefully sharpened the radial furrows with either a mill bill or a mill pick.

After the grain was ground into flour, a small brush or sweeper carried it around the circumference of the stones to a meal spout. From here the flour descended to the floor below for sifting and bagging. Wrote Oliver Evans: "The effect we wish to produce by sifting, or bolting, is to separate the different qualities of flour from each other, and from the skin, shorts, or bran. . . ." Grades of flour were separated by rotating bolting cylinders fitted

Shepherd's Mill operated from about 1739 until 1939. In about 1880, a frame third story was added to the original two-story stone mill. The building, which features an impressive 40-foot iron water wheel, is privately owned.

with cloth meshes of various fineness. These cloth meshes worked much like sieves, and separated the flour into shorts, or bran, carnel, middlings, and superfine flour, respectively.

A miller indulged in few cash transactions during the 1700s and 1800s. Instead, he collected from his customers a toll of from 1/6 to 1/8 of every bushel of wheat, corn, or rye he ground for them. A dishonest miller, however, was often tempted to abuse this system. In Maryland, for instance, laws "for the prevention of the abuse frequently committed by persons keeping water mills, by taking excessive toll. . ." were enacted as early as 1704. But a miller of good character often had a wide influence in his community. At his mill there might take place public sales, horsetrading, and other daily gatherings of social and economic importance to the community. Nor did an enterprising miller just grind wheat, corn, or rye; he might also merchandise beef steaks, veal, potatoes, turnips, butter, molasses, coffee, lumber, firewood, tobacco, matches, flannel, and the like.

Shepherd's Mill operated for about 200 years, closing finally in 1939. About three miles downstream from Harpers Ferry, at the foot of South Mountain in Maryland, the Weverton Grist Mill was erected in 1774. In Millwood, Va., about 25 miles south of Harpers Ferry in the Shenandoah Valley, the Burwell-Morgan Gristmill was built between 1782-1785. Owners Nathaniel Burwell and Daniel Morgan, veterans of the American Revolution, added a dwelling house, tanyard, tanner's house, and a distillery on their property in just a few short years. The Burwell-Morgan Gristmill operated continuously until 1943. In Washington County, Md., just across the Potomac River from Harpers Ferry, a great variety of mills were built in the late 1700s. Along Antietam Creek and Little Antietam Creek, for instance, Dennis Griffith's 1795 Map of Maryland listed 14 gristmills and three forges, all powered by water.

Throughout the Colonial period small local enterprises like Shepherd's Mill and the Burwell-Morgan Gristmill were established as communities grew in size. Many such early settlements, as well as supporting a gristmill, also often had a sawmill where farmers and builders could have lumber cut, and a fulling mill where home-spun cloth could be scoured, cleaned, and dyed. These water-powered mills were relatively small in size, prompted by local needs, and dependent upon local resources: grain, timber, and sheep's wool, for example. Self-sufficiency indeed characterized Colonial mills and the small communities they served.

In time, settlements grew. Demand for manufactured goods increased and so local industries became more diverse. Shoemakers, clothiers, carpenters, blacksmiths, wheel-wrights, and other skilled craftsmen appeared in growing numbers. Forges and furnaces were erected to produce basic iron products for the community: household utensils, pots and pans, tools, stoves, nails, and reinforcement bands for wagon wheels, millstones, and water wheels. As more homes and shops were built, rolling and slitting mills were added to forges to manufacture nails and spikes in large quantities. Brickyards, shingle factories, and furniture shops as well accompanied local expansion.

Still, these many industries were small in scale, dependent upon nearby resources, and isolated by limited transportation facilities. Only as roads and canals improved access to rural areas did local industries become more regional in character.

Antietam Iron Works was one such community industry that participated in this transition into a regional enterprise. Called Antietam Forge on Dennis Griffith's 1795 Map of Maryland, the first furnace stack was erected in about 1765 alongside Antietam Creek less than a mile above its confluence with the Potomac. Harpers Ferry was located about nine miles downstream. Characteristic of the Colonial period, the necessary raw materials for iron production were procured locally. The surrounding property, known as "Indian Deed," provided sizable stands of timber for charcoal, limestone for flux, and water power from Antietam Creek. The only ingredient lacking in the immediate location was iron ore. This, fortunately, was available in abundance along a fault line separating *Harpers shale* and Shenandoah Valley limestones about four miles down the Potomac River. These deposits were mined in open quarries on opposite sides of the river, called respectively the "Maryland bank" and the "Virginia bank." The ore was then transported upriver by flatboat to Antietam Iron Works.

The furnace stack, which was built into the side of a hill, was stoked from the top with about 15 bushels of charcoal, 200-300 pounds of iron ore, and several shovelsful of limestone. A blast of air was provided at the bottom of the furnace by a water-powered bellows. While the furnace was in blast, the charcoal melted the ore into molten iron which collected in a crucible at the base of the stack. The charcoal also combined with oxygen present in the iron ore, burning off as carbon dioxide and carbon monoxide. Calcium carbonate contained in the limestone fluxed out the remaining impurities in the iron ore as slag.

Typically, the furnace might remain in blast for three days. All the while "fillers" kept stoking the furnace at half-hour intervals day and night, consuming about one acre of mature hardwood timber per day. Periodically, perhaps twice a day or more,

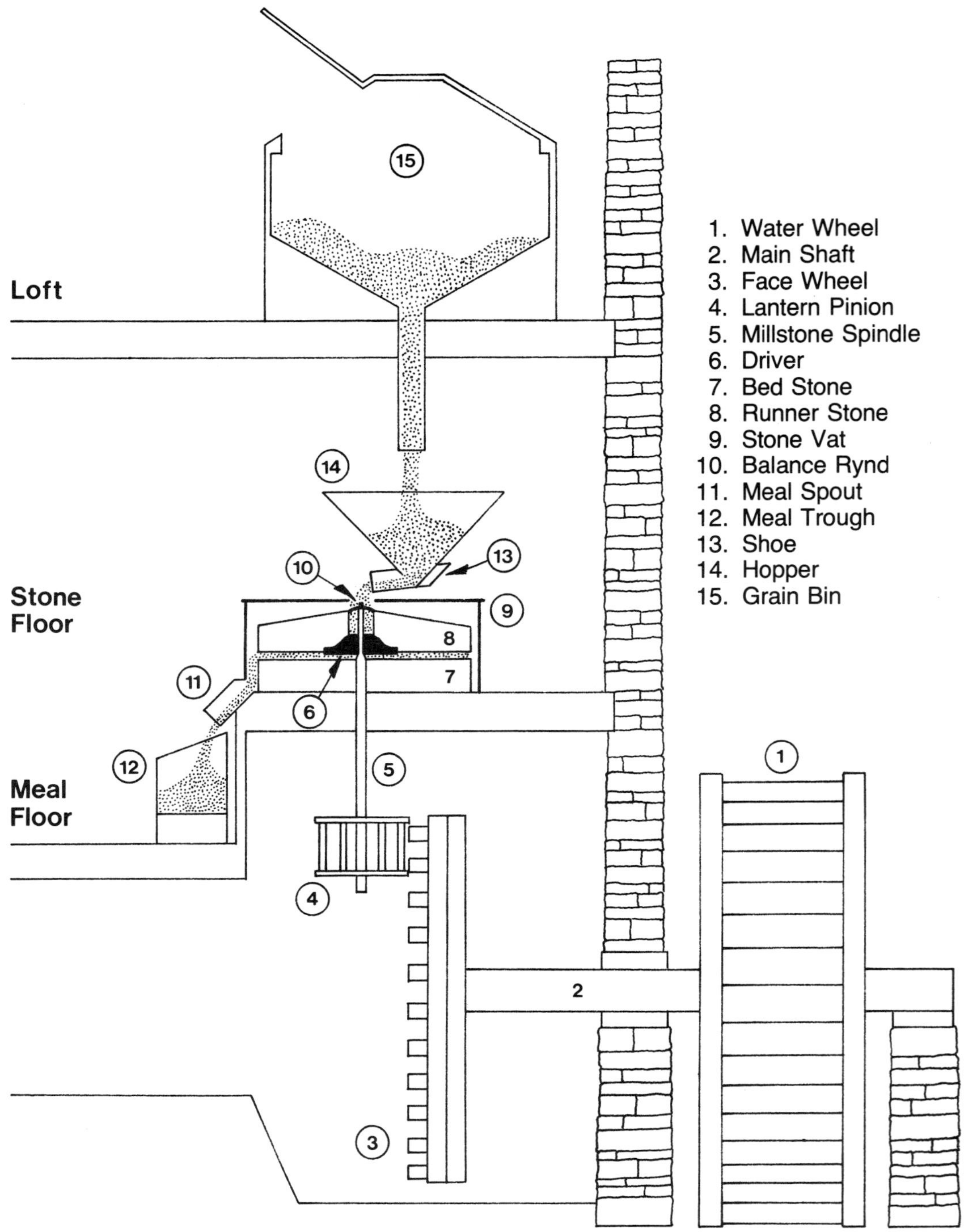

Gristmill Diagram

the "founder" tapped the furnace, draining off the molten iron and the lighter slag residue from the crucible. If the molten iron was of the proper consistency, workers ladled the hot liquid into special sand casting molds. In this manner pots, pans, tools, and stove components were produced.

If the molten iron, however, was of inferior quality, it was tapped off into sand trenches as pig iron. These brittle pig bars were then taken to the adjacent forge and worked with a heavy water-powered trip hammer. Continued heating and hammering drove residual impurities from the pig iron and produced strong, durable wrought iron. Some of this wrought iron was sold to country blacksmiths who turned it into finished goods—tools, implements, reinforcement bands, etc.—for farmers, wagoners, and mill owners. The rest was consigned to the rolling and slitting mill at Antietam Iron Works. Here, with power provided by a 14-foot overshot water wheel, sheet iron and nail rods were made. At the nailery, which in 1841 was equipped with 19 nail and spike machines, nail rods were cut into nails and spikes. A shingle, jointing, and stave mill, located next to a sawmill, manufactured kegs for these nails. A two-story frame gristmill, containing four run of fine French burr-stones and powered by two 17-foot water wheels, stood at the very end of the long headrace conveying water from Antietam Creek. In all, by about 1841, water from this headrace powered nine separate water wheels at Antietam Iron Works. The total "head" and "fall" was about 21 feet.

A stable and company store also stood on the property. Workers, who generally labored for 12-hour shifts, were paid in part with credit notations from the company store. Here they could purchase virtually all their supplies: food, clothing, liquor, and most household goods.

Antietam Iron Works was a genuinely self-sufficient enterprise. Not only were the raw materials for iron manufacture obtained locally. Virtually all the needs of the community were available in the area or produced on site: stone for building material, water for power, lumber from the sawmill, flour from the gristmill, iron tools and utensils, and all kinds of other iron products for builders, mill owners, farmers, and wagoners.

In time, improvements in transportation rendered distant markets more accessible. The Patowmack Canal improved river navigation downstream to Georgetown and Alexandria after 1785. On the C & O Canal, completed here in 1834, pig iron was boated to the Harpers Ferry Armory. At about this time, Antietam Iron Works was produc-

Gearing arrangements often varied from one gristmill to another. At the Burwell-Morgan Gristmill in Millwood, Va., a single face wheel drives two spur wheels, two lantern pinions, and two pair of burr-stones. One of these lantern pinion gears and a spur wheel behind it are pictured here.

A wooden hopper is mounted on top of a stone vat inside Peirce Mill. The mill, located in Rock Creek Park in Washington, D.C., was built in 1820, repaired as a WPA project during the Depression, and restored by the National Park Service in the 1960s.

In this picture, the runner stone has been lifted off the bed stone by an iron stone crane and set on its side for re-dressing. A cast-iron driver is attached to the top of the millstone spindle, which protrudes from the center of the bed stone.

ing 1,476 tons of pig iron annually. Nails, which sold for 25¢ per pound nationally before 1800, cost as little as 3¢ a pound in 1843 due to continued improvements in water-powered machinery. At Antietam Iron Works alone, 400-500 kegs of nails were produced per week after 1841.

Industry here was indeed expanding beyond the limits of local community needs and taking on all the characteristics of a regional enterprise. But limits to growth in this particular instance were soon reached. Newer furnaces burning anthracite coal with heated air blasts produced better quality iron faster and cheaper. Antietam Iron Works, located far from coal supplies on the one hand, large urban markets on the other, and lacking railway connections to either, began to fade. When the iron ore deposits were finally depleted from the "Maryland bank" and the "Virginia bank," the end had come. In 1886, Antietam Iron Works closed down for good.

Other iron manufactories in the Harpers Ferry area were also founded upon the needs of the local community and the availability of nearby resources. On May 10, 1742, William Vestal engaged a builder to construct a "Bloomery for making Barr Iron, upon the present plantation of William Vestal lying upon the Shunandore." A "Bloomery" forge produced only limited amounts of wrought iron using primitive methods—the iron ore was heated with charcoal into a pasty state, then placed under a hammer and worked until many of the impurities were driven off. William Vestal's "Bloomery" was located in the present vicinity of Keyes Ferry, about five miles up the Shenandoah River from Harpers Ferry. Keep Triest Furnace was erected in about 1764 near the mouth of Elk Run, about two miles up the Potomac River from Harpers Ferry. Smaller in size than Antietam Iron Works, this establishment also obtained iron ore from the nearby "Maryland bank" and "Virginia bank."

To fully comprehend the establishment of these various water-powered mills and the development of industry around Harpers Ferry, we must explore the properties of falling water. Again, "head" and "fall" together denote the difference in height between supply water in a mill's headrace and waste water in its tailrace. "Head" is all the water above the point of impact on a water wheel; "fall" is all that water below the point of impact. The effectual descent of water at a particular mill-seat, however—a measure of the potential energy falling water can produce—is comprised only of half the "head" added to the whole "fall." This is because falling water issuing from a headrace imparts only half its force by impulse in striking a water wheel. But once contained in the buckets or floats of the wheel, water by its gravity expends its full energy through the "fall." Consequently, "the whole descent of the water of a mill-seat, should be nicely divided between head and fall, to suit each other, in order to obtain the best effect, and a steady-moving mill." So wrote Oliver Evans in *The Young Millwright and Miller's Guide.*

The type of water wheel a millwright chose, then, was of some consequence. Logically, the less the "head" and the greater the "fall," the more efficient would be the wheel in converting falling water into motive power. Also, depending upon the volume of water supplied by the headrace, the millwright had to determine the wheel's proper width. Since small streams usually had a high rate of descent, tall, narrow water wheels were used here. On the other hand, streams and rivers with large volumes of water often had a lesser descent; here wide wheels with a small diameter were used. Another consideration was the type of machinery the water wheel was intended to drive. Gristmills, sawmills, and forges, for example, all had different power requirements. Using basic principles of math and physics, the millwright thus designed his wheel with care and precision.

Prior to the 1840s, a millwright could choose from four basic types of water wheels: undershot, breast, pitch-back, and overshot.

Undershot water wheels were generally used for falls of two to nine feet (see diagram, page 22). In this particular construction, the effectual descent of the water was comprised of all "head" and zero "fall." Wrote Oliver Evans:

> *Undershot wheels move by the percussion or stroke of the water, and are only half as powerful as other wheels that are moved by the gravity of the water. Therefore, this construction ought not to be adopted except where there is but little fall, or great plenty of water.*

Undershot wheels were only about 30% efficient in converting the energy available in falling water into motive power.

Breast wheels were appropriate for falls of six feet upwards of 25 feet (see diagram, page 24). In this arrangement "head" and "fall" were roughly equal. The name breast wheel derived from the height at which water struck the wheel; as this height varied from mill-seat to mill-seat, low breast, middle breast, or high breast constructions might be used. Like undershot wheels, water passed under a breast wheel. But in addition deep buckets or floats were set into slots in the wheel rims and held the water on the wheel through the "fall." Thus a breast wheel was turned both by the impulse of the water striking the wheel and by the weight of the water contained in the buckets. A stone apron built close to the wheel held

The ruins of the Weverton Grist Mill are adjacent to Lock 31 on the C&O Canal. The original mill was torn down and replaced with a three-story frame structure in 1856. This particular stone arch admitted water from the canal into the mill and provided a head and fall of 17 feet.

Ruins of a single furnace stack at Antietam Iron Works. The outer surface was constructed of stone, while the interior chamber was lined with heat-resistant brick. By 1833, this single stack was producing 1,476 tons of pig iron per year.

this water in the buckets until it was discharged into the tailrace.

Pitch-back wheels operated much like breast wheels. The only difference was that supply water, rather than striking the wheel at breast height, was instead admitted onto the wheel very near the top. The water still passed under the wheel, and an apron still held the water in the buckets. But because the water expended its energy in the form of gravity through a much longer "fall," pitch-back wheels were more efficient than breast wheels.

Overshot wheels were preferred by millwrights for falls over 10 feet. Like breast wheels and pitch-back wheels, deep buckets set into slots in the wheel rims held the water on an overshot wheel. But in the overshot arrangement water passed over the top of the wheel before falling into the tailrace. Overshot wheels were moved almost entirely by the weight of the water exerted through its "fall"; the impulse of the water issuing from the headrace contributed only slightly. This construction produced a water wheel about 60% efficient in converting the energy of the falling water into motive power.

Oliver Evans, in *The Young Mill-wright and Miller's Guide,* provided a simple equation to help millwrights compare the performances of these various water wheels at a particular mill-seat. First,

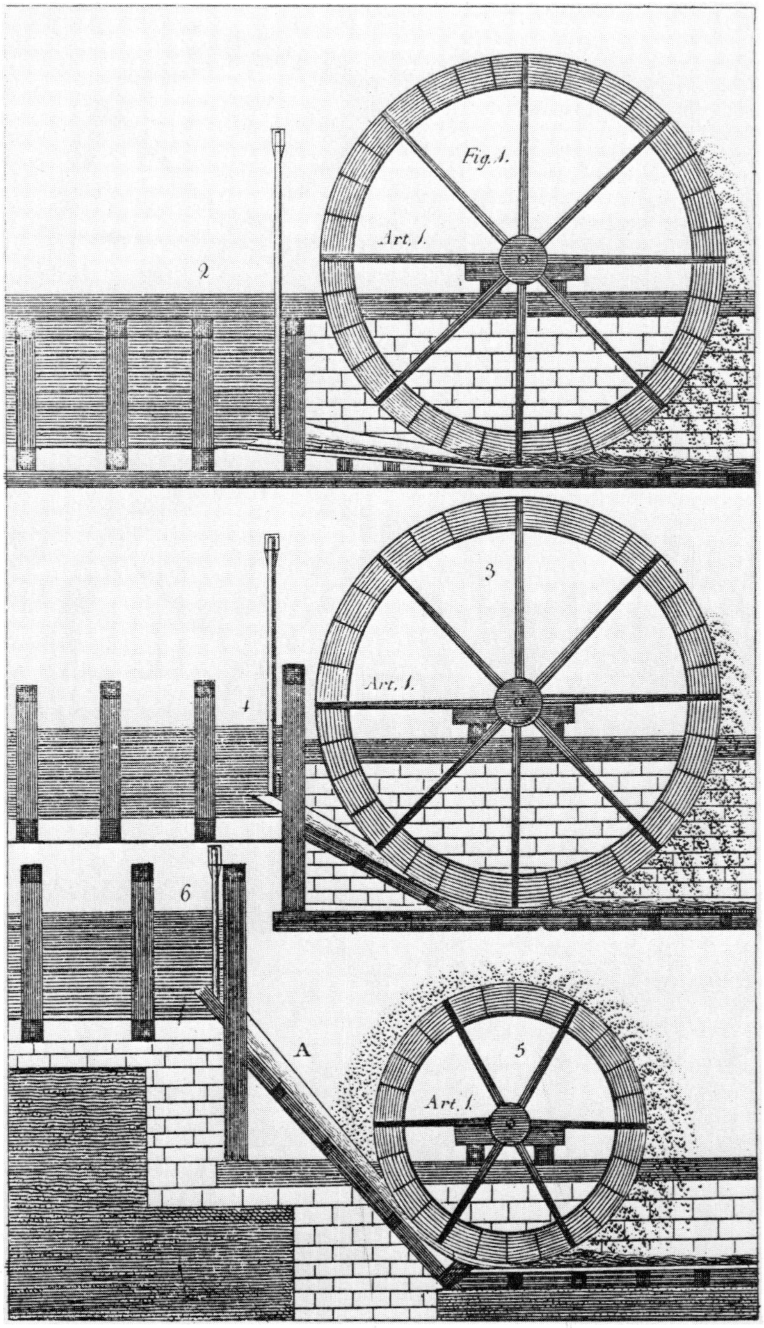

Illustrations of three undershot water wheels, from The Young Mill-wright and Miller's Guide, *by Oliver Evans (13th edition, published in 1850).*

if you multiply one cubic foot of water expended from a headrace by one foot of effectual descent—half the "head" plus all the "fall"—you obtain one unit of power. Now apply this equation to a mill-seat where supply water, for instance, amounts to 16 cubic feet per second (cfs) and the vertical descent is 16 feet. The following table compares the results obtained for three different water wheels:

Type of Wheel	Head in ft.	Fall in ft.	Effectual Descent	×	Supply Water (cfs)	=	Units of Power
Undershot	16	0	8	×	16	=	128
High Breast	6	10	13	×	16	=	208
Overshot	2	14	15	×	16	=	240

While Evans' equation does not consider such factors as width of water wheel and fluctuations in streamflow, still this table compares the relative performances of these three types of water wheels. As well, it illustrates the important relationship between falling water and a millwright's choice of water wheel.

Shepherd's Mill and those that soon followed established the early use of water power in the Harpers Ferry area. Two reasons were the presence of ample falling water and the know-how of millwrights to harness it. In addition, the local scope of community living and the availability of nearby resources contributed enormously to the proliferation of these small mills in the late 1700s. That this proliferation of mills should spread to Harpers Ferry was inevitable. Here abundant supplies of water power and building materials, and, in the coming years, improving facilities for transportation were all concentrated at one site.

For Harpers Ferry, the 19th century held great promise.

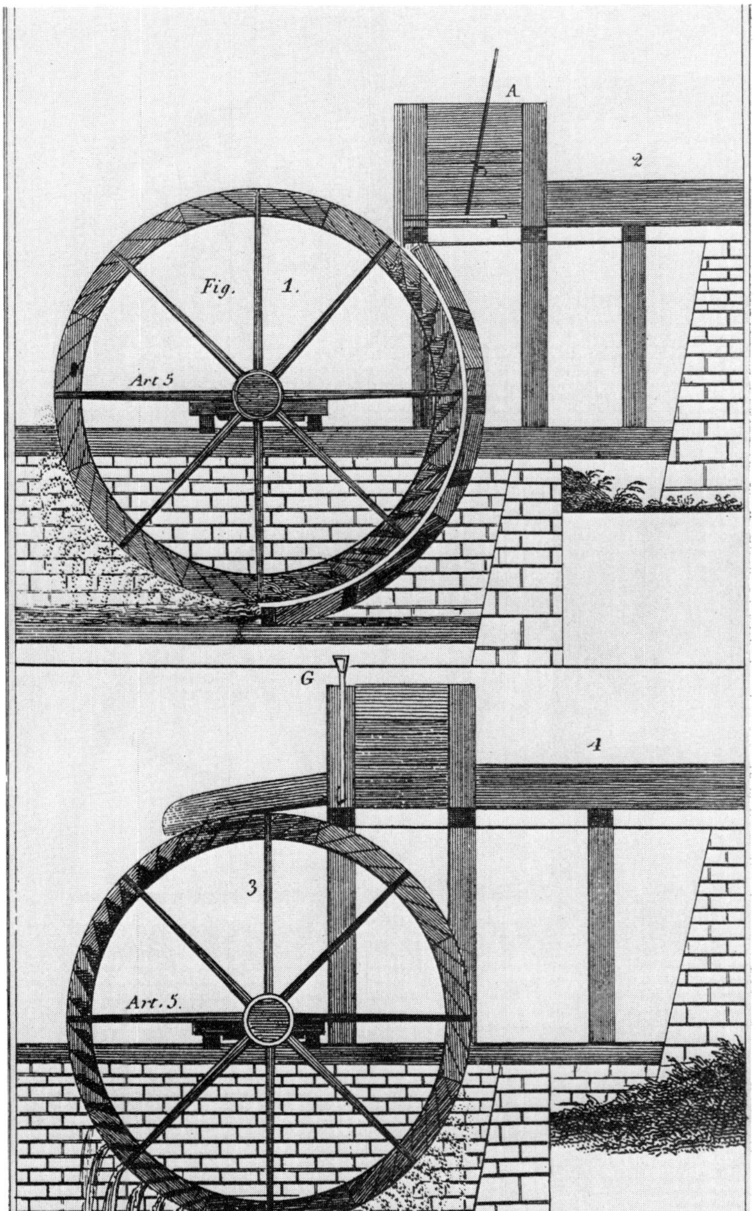

Illustrations of a pitch-back wheel (top) and an overshot water wheel, from The Young Mill-wright and Miller's Guide, *by Oliver Evans (13th edition, published in 1850). Wheel rims were usually built of segments, or felloes, of oak. Rot-resistant elm was preferred for the buckets or floats of the water wheel.*

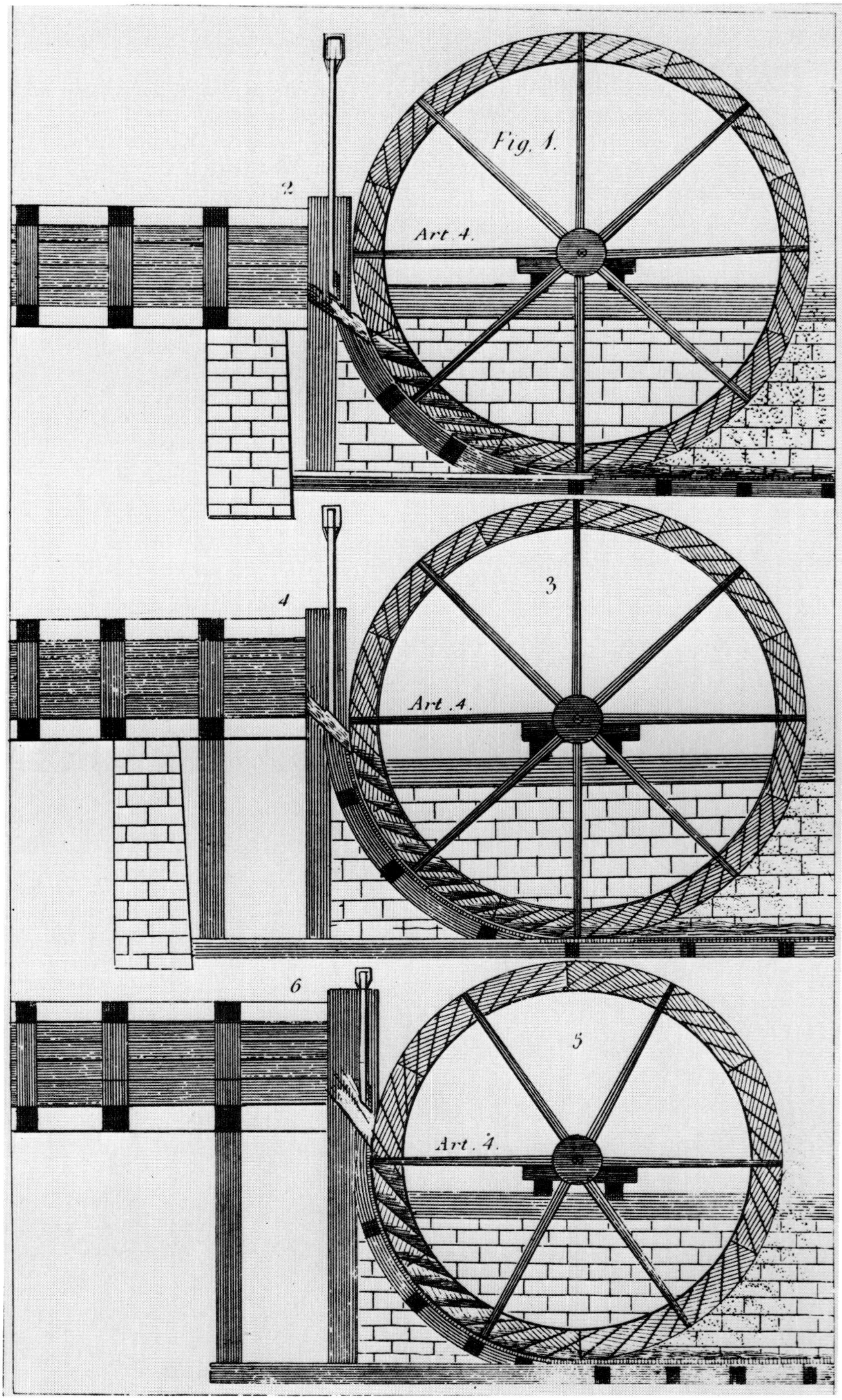

Illustrations of a low, middle, and high breast wheel, from The Young Mill-wright and Miller's Guide, *by Oliver Evans (8th edition, published in 1834). The protruding ends of iron gudgeons, which were inserted into the ends of the wheel shafts, normally turned on wood, stone, or brass bearings.* National Park Service

Chapter Four

LOCATING THE ARMORY AT HARPERS FERRY

On June 15, 1796, the federal government purchased an 118-acre parcel of land at Harpers Ferry from John Wager, Sr. This purchase, concluded by Tobias Lear of Georgetown and George Gilpin of Alexandria, was the end result of recent congressional legislation and much controversy within the government. Three years later, construction commenced on the "United States Armory and Arsenal at Harpers Ferry."

The establishment of this manufactory had its origins in America's struggle for independence. During the American Revolution, the flow of arms to the Colonies from European exporters was seriously disrupted. Domestically, both the supply and quality of weapons manufactured by private firms was unreliable. As a consequence, the federal government decided during the years after the war to establish and oversee its own arms maufactories. In April 1794, President George Washington sponsored a bill "for the erecting and repairing of Arsenals and Magazines." The legislation, which gave the president substantial latitude in choosing where to build these facilities, was quickly approved by Congress.

Without complaint, Springfield, Mass., was selected as the location of the first national armory. President Washington then determined to consolidate the remaining appropriations provided by the legislation and erect a second new armory and arsenal at Harpers Ferry.

In choosing the Harpers Ferry site, Washington claimed that the place possessed an "inexhaustible supply of water" for power and "centrality among Furnaces and Forges." Washington also cited an abundance in the area of timber for charcoal to fuel the forges of the new armory. Less apparent in the president's decision, but of equal importance, was his regional knowledge of the whole Potomac River valley.

Born at Wakefield on the south shore of the Potomac estuary in 1732, George Washington spent most of his childhood and adult years along the banks of this river. As a young surveyor for Lord Fairfax, he had occasion to travel across much of the great Northern Neck estate. In particular, Washington accompanied James Glenn and George William Fairfax on land surveys in the lower Shenandoah Valley around Harpers Ferry during March 1748. Later that same spring he travelled up the Potomac River to Cresap's (Oldtown) and helped run several surveys in the remote South Branch Valley. These trips not only served to familiarize the 16-year-old Washington with the geography of the region, but brought to his attention the many resources of the area as well.

Knowledge of the region gained from these and subsequent travels—particularly during the French and Indian War—prompted Washington to lobby for the creation of the Patowmack Company. This enterprise, which was duly formed in 1785, promised to render the Potomac River navigable upstream to Westernport, Md. Washington, appropriately, was elected the company's first president, and in August 1785 he conducted a tour of the many rapids the Patowmack Canal would have to surmount. During this trip Washington had occasion to spend a night at the Harper House while he studied the rapids around Harpers Ferry. He referred to this accommodation as "The Tavern;" since Robert Harper's death three years earlier, John Wager, Sr., had leased the building as the town's only inn.

Washington's insistence on the Harpers Ferry location for a new armory and arsenal was surely a product of his personal familiarity with the site. In a regional context, the president's decision reflected as well a basic desire to further the economic development of the entire Potomac River valley. Washington had already made strides in this direction when he played an active role in the affairs of the Patowmack Company. That he chose Tobias Lear and George Gilpin, both prosperous merchants and stockholders in the Patowmack Company, to act as his special agents in purchasing the Harpers Ferry property further demonstrated the ties between the president and the economic fabric of the area.

Ultimately, Washington's regional concerns in selecting the Harpers Ferry location were a product of his 1790 decision to build the new national capital at the head of the Potomac estuary. The president surely concluded from his wide knowledge of the area that a vigorous regional economy linked together by dependable river-borne commerce would help insure the prosperity of his federal city. The construction of a new armory and arsenal was an important component of this overall campaign.

The War Department, however, did not view the Harpers Ferry decision with the same vigor as the president. Secretary of War Henry Knox, who favored the Springfield location, had then voiced his preference for expanding three small establishments rather than building a single large one.

The new arsenal dominates this 1803 print of Harpers Ferry. The Harper House, which George Washington referred to as "The Tavern" in 1785, stands on the side of the hill to the left of the arsenal. The Potomac Ferry is pictured crossing the river. National Park Service

Knox's successor, Timothy Pickering, argued that appropriations by Congress "would be inadequate for a single new establishment." What's more, a study commissioned by the War Department during the spring of 1795 did not even mention Harpers Ferry as a suitable site along the Potomac River. The report's author explained that "there was no ground on which convenient buildings could be placed" here and that "no water work would be safe" from exposure to floods.

The president nevertheless stood his ground. In a letter to Secretary Pickering in September 1795, Washington again endorsed the Harpers Ferry location, and both Lear and Gilpin proceeded with the necessary property acquisitions.

The 118-acre parcel of land purchased from John Wager, Sr., was bordered by the Potomac and Shenandoah rivers and by a line running between both rivers along what is now Union Street. John Wager, Sr., was married to Sarah Harper Wager, Robert Harper's niece and heir to his original 125-acre property. Although neither Wager nor his wife ever lived in Harpers Ferry—their home was in Philadelphia—they nevertheless retained two key parcels when they sold land to the government. The "Six-Acre Reservation" was comprised of the present land along Shenandoah Street between Potomac Street and the Stage Coach Inn. The three-quarter-acre "Ferry Lot Reservation" was situated at the confluence of the Potomac and Shenandoah rivers.

John Wager, Jr., and his wife Catherine, who inherited both parcels together with the ferry concession and moved to Harpers Ferry in 1803, exacted the utmost commercial advantage from these centrally located lots. As the armory's workforce expanded, the Wager family was able to monopolize virtually all mercantile transactions here; food, clothing, and liquor were available exclusively from establishments on the "Six-Acre Reservation" and the "Ferry Lot Reservation." According to Catherine Wager, this arrangement was a formal one. In a letter to Secretary of War John C. Calhoun dated Sept. 9, 1818, she claimed that Tobias Lear had agreed when he purchased land for the government "that the Person living on her land shall have a monopoly of the mercantile Business at the Place, & that the Persons employed about the Armory, shall be constrained to take all supplies from the Persons occupying her land and tenements at the Ferry."

The Wager family's monopoly endured at Harpers Ferry until about 1836. Even as late as 1860, Wager family descendents still owned about one-third of all the lots on the "Six-Acre Reservation" and the "Ferry Lot Reservation." The high cost of mercantile goods brought about by this situation contributed significantly to the high wages necessarily paid to armory workmen. This in turn forced up the cost of arms production, a situation that the War Department found both frustrating and irritating right up to the Civil War.

Construction of the first armory buildings and waterworks began in the spring of 1799. By the end of that year, the smith shop and the factory were completed along the Potomac shoreline. The smith shop was a two-story brick workshop, 80 feet by 26 feet. On the first floor there were 10 hearths for forging small gun components. The second floor contained benches for lock filers. The factory was a two-story brick finishing shop, 120 feet by 26 feet. Water-powered boring, grinding, and polishing machines were located on the first floor, while the second floor had benches for stocking and assembling the arms. The mill, where heavy forging was done, was completed around the end of 1800. Here a heavy tilt-hammer for working iron and preparing gun barrels was powered by a breast wheel eight feet in diameter. At the time, this was the largest of the five water wheels located on the armory grounds. The mill also contained forges for welding the gun barrels.

The arsenal, a three-story brick building 120 feet by 27 feet, was just about finished in December 1799. This structure was located near the confluence of the Potomac and Shenandoah rivers, adjacent to the "Ferry Lot Reservation." Here arms manufactured in the armory shops were stored.

Construction of the waterworks for the Harpers Ferry Armory took more time. On Aug. 9, 1799, work began on the Armory Dam (known also as the U.S. Potomac Dam, Armory Potomac Dam, and, in C&O Canal Company records, as Dam No. 3). In just 64 days, a masonry foundation was completed. But the superstructure of the dam, which was comprised of square timber piers bolted together and filled with stone, was not finished until late 1800. Excavation of the Armory Canal (or the U.S. Armory Potomac Canal) also proceeded slowly. Due to an inadequate supply both of skilled and unskilled labor, the one and one-quarter-mile-long headrace was not completed until 1801.

Because the various forging, grinding, boring, and polishing machines used at the Harpers Ferry Armory relied on water power to operate, delays in completion of the waterworks prevented the start of full-scale arms manufacture for many months. Even after completion, government records indicate that the Armory Canal leaked so badly that heavy forging work could not be effectively performed.

Nevertheless, in 1801 limited production of new arms commenced with the completion of 293 muskets. The following year full-scale arms manufacture began; a work force of 25 men produced 1,472 muskets. By 1807, the number of armory workmen had increased to 67. In 1810, the Harpers Ferry Armory included 12 workshops, 15 dwelling houses, and a work force of 197 men. Annual production of muskets now averaged about 10,000; "short" rifles and pistols were manufactured in much smaller quantities.

At about this time musket manufacture was divided into four categories: lock, mounting, stock, and barrel. The lock, or firing mechanism, was the most fragile and complicated part of the musket. The several lock components—lock plate, tumbler, bridle, sear, hammer, pins and screws—were carefully produced with forging, drilling, and filing tools. The mounting, which was comprised of trigger guards, metal bands, and heel plates, was made with the same precision machinery and craft skills.

The shaping of the wooden gun stock was basically a hand operation in the early 1800s, requiring a single craftsman to whittle, bore, and chisel each stock. This process was mechanized, however, in 1818, when Thomas Blanchard introduced a lathe for turning irregular forms. The following year this new machine was installed at the Harpers Ferry Armory. Blanchard's lathe, which in subsequent years was developed into 16 separate units, used a standard gun stock as a template from which identical copies were cut.

Producing the gun barrel was the most expensive and physically demanding part of musket manufacture. First a piece of bar iron was rolled into a flat, long, and narrow "skelp." At the barrel forging shop this "skelp" was welded by hand into a barrel. In 1840, water-powered trip-hammers replaced manual methods of welding gun barrels at Harpers Ferry, producing a sounder seam in about half the time previously required. Annealing restored malleability to the gun barrel, which was rendered brittle by several heatings during welding. Next the barrel was straightened by hand and the bore was drilled. In the proof house the barrel was charged with powder and ball to test the strength of the weld. A barrel that withstood two charges was given a proof mark and sent back to the boring mill to be finished bored and polished. Finally, the barrel went to the finishing shop for final assembly with the lock, mounting and stock.

John Edwards Caldwell, in *A Tour Through Part of Virginia in the Summer of 1808*, wrote that at the Harpers Ferry Armory "the weighty part of business is conducted by the aid of machinery turned by water. . . ." Still, the smaller components of a well-made musket were produced primarily with the skillful and dexterous use of hand tools. A typical workman could not simply be hired and put to work making muskets. Rather, he had to acquire the necessary craft skills by serving as an apprentice under a master craftsman. Indeed, between

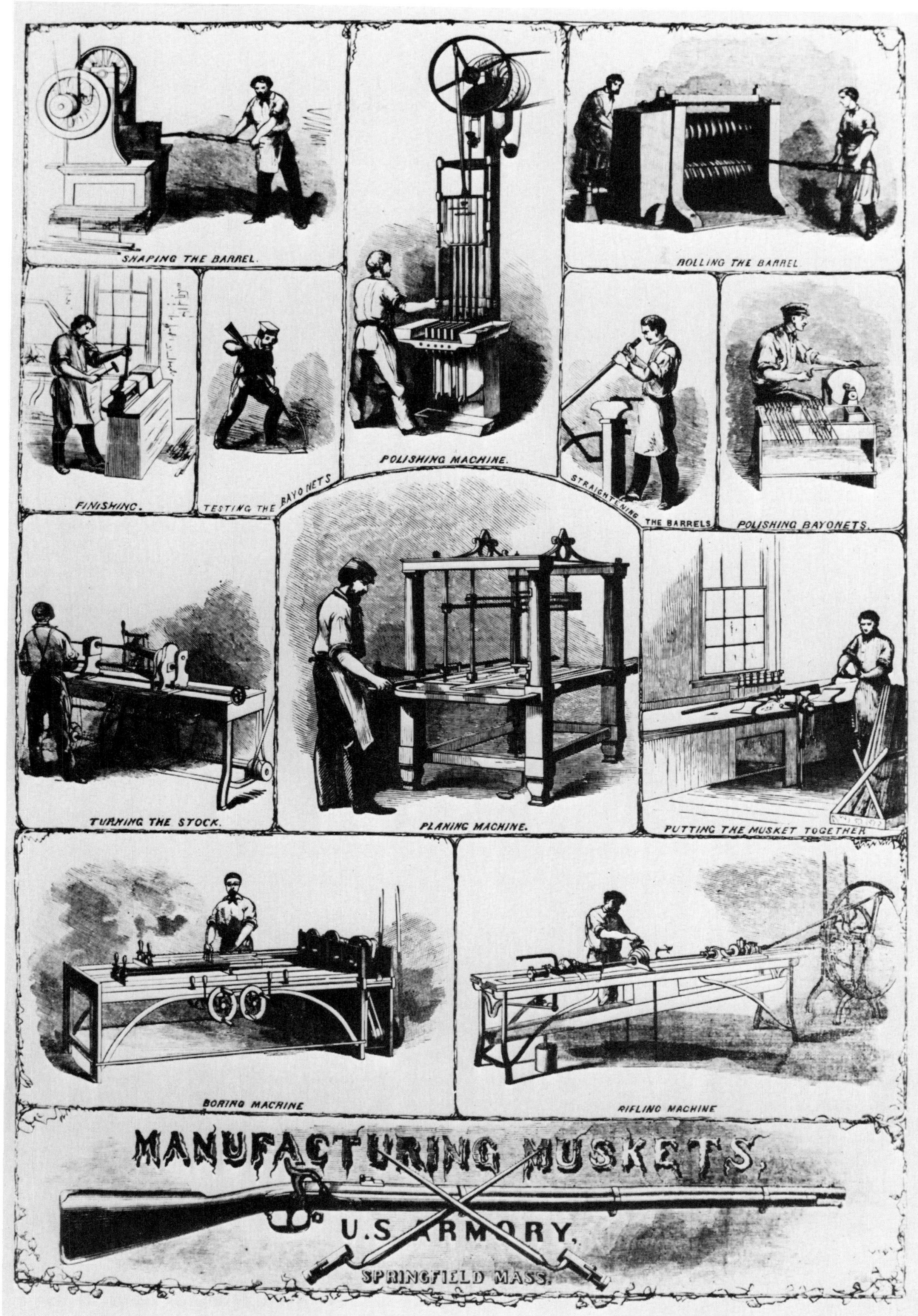

These illustrations of arms manufacture at the Springfield Armory appeared in Harper's Weekly *on Sept. 21, 1861.*

National Park Service

1800-1813, some 30 boys served such apprenticeships at the Harpers Ferry Armory. The term of apprenticeship was necessarily long—from five to seven years—since individuals had to master the production of a variety of precise components with hand tools rather than just one particular item in assembly-line fashion. A young man, after he completed this thorough training, then gained experience by working as a regular armory workman or as a journeyman gunsmith elsewhere. Finally, he finished his career as a master craftsman.

The apprenticeship system became less and less necessary as specialized water-powered machinery, such as Blanchard's lathe, was introduced. Individual workmen, here as elsewhere, were increasingly being trained only to operate precision machines and to produce individual components. This specialization laid the groundwork for a factory system of production, characterized both by mechanization and by a division of labor. Gradually, wage-earning laborers operating according to a strict daily routine began replacing skilled craftsmen who to a large extent had determined their own pace of work. Still, as late as 1840, workmen at the Harpers Ferry Armory continued to exercise a large measure of manual dexterity and individual craft talent in daily arms manufacture.

The establishment and subsequent growth of the U.S. Armory at Harpers Ferry contributed to the establishment of other local industries as well. In 1824, Lewis Wernwag erected a large frame saw mill on the "Island in the Shenandoah River near Harpers Ferry." This island, which contained about 13 acres, was commonly called Stubblefield's Island. The name derived from armory superintendent James Stubblefield, who a year earlier had acquired the property. In 1827, by an act of the Virginia General Assembly, the town of Virginius was established here. The property subsequently became known as the Island of Virginius, or simply Virginius Island.

That the "head" and "fall" available along the Shenandoah River here was 14 feet did not go unnoticed by Lewis Wernwag and other local millwrights. Wernwag's saw mill was probably first fitted with a water-powered reciprocating up and down

This former armorer's dwelling house, which still stands along Shenandoah Street, was built between 1822-1830 and substantially enlarged in the 1840s. The building was a government-owned residence which was rented out to armory workmen and their families.

1857 lithograph of Virginius Island. The large building in the center of the picture is the "Island Mill" or Herr's Mill. The four-story building at the far left is a cotton factory, which was erected in 1847 (see Chapter 6). By 1859, there were 37 buildings on the island and a population of almost 200.

National Park Service

saw, geared to cut at 120 strokes a minute. A carriage, also activated by the water wheel, drew individual logs to meet the saw and back again for as many cuts as were desired. More than 20 years later, on Sept. 3, 1846, an advertisement in the *Virginia Free Press* noted that the saw mill now had two reciprocating saws of Crosbey's patent, two circular saws each 24 inches in diameter, and two carriages with their attendant straps.

Lewis Wernwag, a bridgebuilder and inventor of some reputation, managed the operation of his saw mill for about nine years. During that time he also engaged in other local construction projects. In 1824, under contract with the Wager family, Wernwag began erecting a "handsome double wooden highway bridge, 750 feet from one abutment to the other," across the Potomac River. This bridge replaced the old Potomac Ferry, which had been operating since Robert Harper's time. Four years later, he rebuilt the Armory Dam for the government at a cost of $5,500.

In 1832, Wernwag built a three-story stone machine shop just a short distance from his saw mill on Virginius Island. This new establishment contained turning lathes, screw plates, taps and dies, bench screws, and mill screws for making or replacing all kinds of machinery. From 1836-44, for example, the machine shop manufactured turning lathes, saws, and a patent washing machine.

In 1833, after operating his saw mill and machine shop for a brief period under the name "Wernwag & Sons," Lewis Wernwag sold his remaining interest in the business to his son John. Lewis Wernwag proceeded in 1836 to erect the timber spans of the first B&O Railroad bridge across the Potomac River—perhaps his most important structure at Harpers Ferry. John, meanwhile, in partnership with Jesse Schofield, operated the saw mill and machine shop under the name "Wernwag & Schofield."

Another water-powered mill was erected on Stubblefield's Island in 1824 by Fontaine Beckham. Called the "Island Mill," this establishment was a flour mill of some size. Beckham described his mill in the *Virginia Free Press* on Aug. 16, 1832:

The building, a substantial one of stone is 60 feet by 40; it has two water wheels, two pair of burrs, and room for another pair. There is also an excellent pair of country stones. It is now undergoing repairs, and will be in complete order in a few days. . . .

In 1838, the firm "Rowland, Heflebower & Company," which had recently been leasing the mill from Fontaine Beckham, agreed to purchase this establishment. Misfortune, however, soon struck these new owners when the "Island Mill," together with 20,000 bushels of wheat, was destroyed by fire in February 1839.

The following year, Luther Cox erected a large new three-and-one-half-story stone flour mill on the very same site. In 1844, ownership of this building passed to Abraham H. Herr and his brother John, who operated the mill under the name "Herr & Brother." In time, the building became known simply as Herr's Mill.

In about 1824, Townsend Beckham—Fontaine's brother—built an oil mill and opened a tanning business near the upstream end of Stubblefield's Island. In 1835, Hugh Gilleece leased this property and began operating an iron foundry in the old oil mill. Five years later, on June 11, 1840, Gilleece announced in the *Virginia Free Press* that he had "completed his Chopping Mill, on the Island of Virginius, near the Foundry, where he is prepared to chop Rye and Corn. . . ."

Hugh Gilleece's iron foundry was of respectable size and capacity, being about 76 feet by 40 feet and shaped like an L. Of the variety of iron products made here, most were destined for agricultural customers. There were threshing machines of various design, straw cutters, windmill irons, ploughs, McCormick Mouldboards, wagon boxes, branding irons, and coal stoves and grates. As water-powered machinery began to play a more prominent part in local industry, Gilleece produced mill gearing and machine castings in ever increasing numbers. In 1842, for instance, the foundry cast an iron shaft for the Harpers Ferry Armory weighing 3,600 pounds. On July 27, 1843, Gilleece placed the following advertisement in the *Virginia Free Press:*

I have made arrangements with Mr. John Wernwag, whose extensive machine shop is convenient to the Foundry, for the turning and fitting up of every description of Castings. His name as a Machinist, is a sufficient guarantee for the manner in which the work shall be done. Mill Spindles made and repaired.

Together, Hugh Gilleece and John Wernwag, using precise patterns and specifications, made a variety of machine castings for the Harpers Ferry Armory during the 1840s.

Indeed, as the government pumped more money into armory expansion, particularly after 1840, Virginius Island grew by leaps and bounds. By 1844, there were 29 buildings here and a population in excess of 113. The saw mill, machine shop, island mill, iron foundry, and chopping mill produced finished goods for local residents—many of them armory workmen and their families—for nearby farmers and, in particular, for the expanding U.S. Armory. Still, these several establishments were not so much factories as workshops for skilled craftsmen. While water-powered machinery performed the lion's share of work—sawing lumber, grinding grain, and pounding out wrought iron—still a great deal of labor required manual dexterity. Logs had to be properly fed to saws, millstones had to be carefully sharpened, and particular talent was needed to forge iron machine castings and farm implements. As well, individual workmen usually performed not one, but a variety of functions within a single establishment.

Only with larger investments of capital, increased mechanization, a more specialized division of labor, and improvements in transportation could factory methods of production be effected. As the following chapters will show, these many developments were already on their way.

Armory Grounds, 1859

1. Winchester & Potomac Railroad depot
2. Gault House Saloon
3. B&O Railroad office
4. Toll house
5. B&O Railroad depot
6. Wager House Hotel
7. Potomac Restaurant
8. Old superintendent's office
9. Large arsenal
10. Small arsenal
11. Fire enginehouse and watchman's office
12. Armory offices
13. "Bell" or finishing shop
14. Polishing shop and wash-house
15. Boring mill
16. Stocking and machine shop
17. Millwright shop
18. Grinding mill, sawmill and carpenter's shop
19. Tilt-hammer and barrel-forging shop
20. Lumber house and coal bin
21. New stock storehouse
22. Warehouse
23. Smith and forging shop
24. Annealing shop and brass foundry
25. Proof house
26. Charcoal house
27. Old stock storehouse
28. Rolling mill

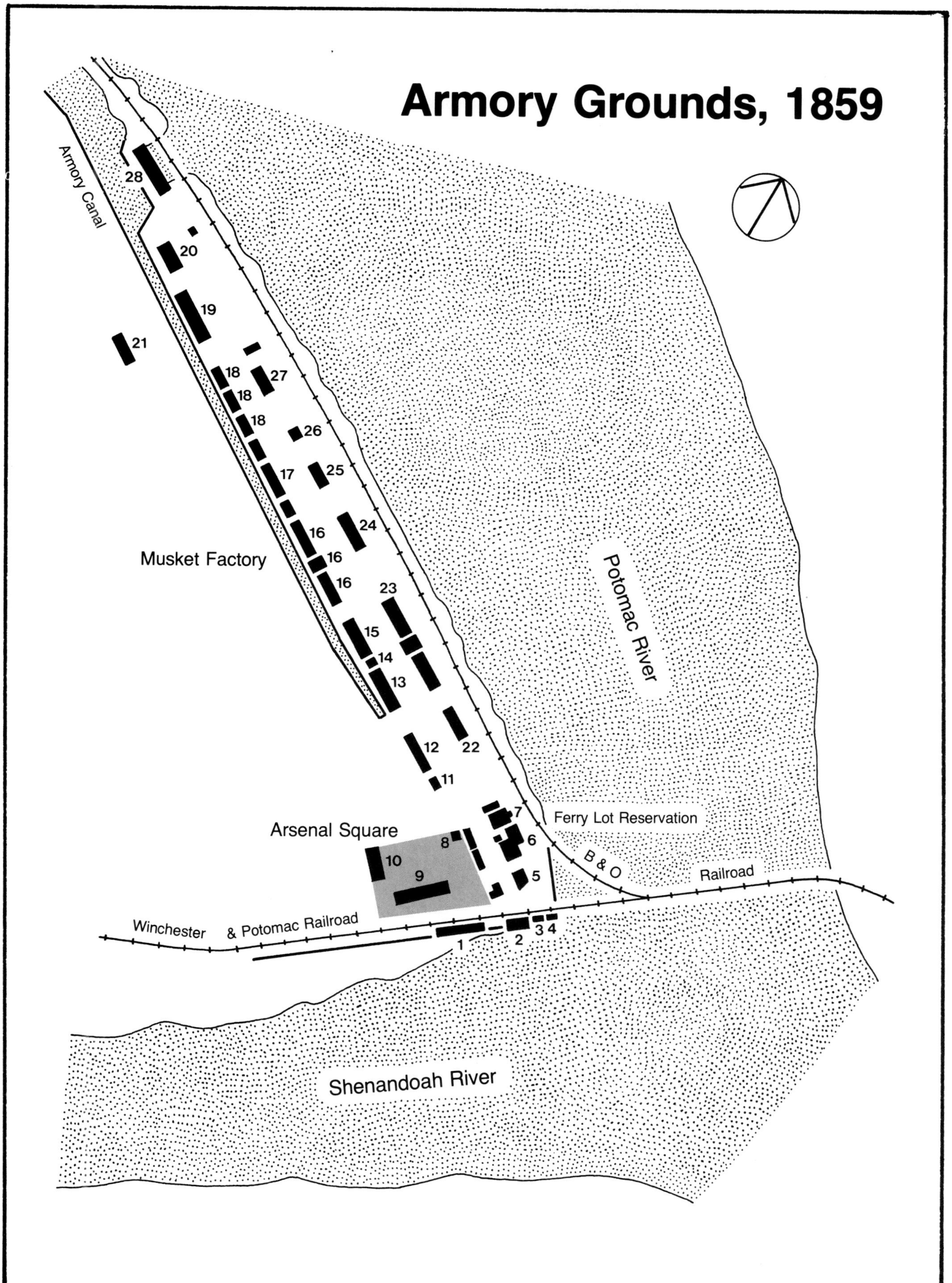
Armory Grounds, 1859
Armory Canal
Musket Factory
Potomac River
Ferry Lot Reservation
Arsenal Square
B & O
Railroad
Winchester
& Potomac Railroad
Shenandoah River
28
20
19
21
18
27
18
18
26
17
25
16
24
16
16
23
15
14
13
12
22
11
7
6
8
10
9
5
1
2
3
4

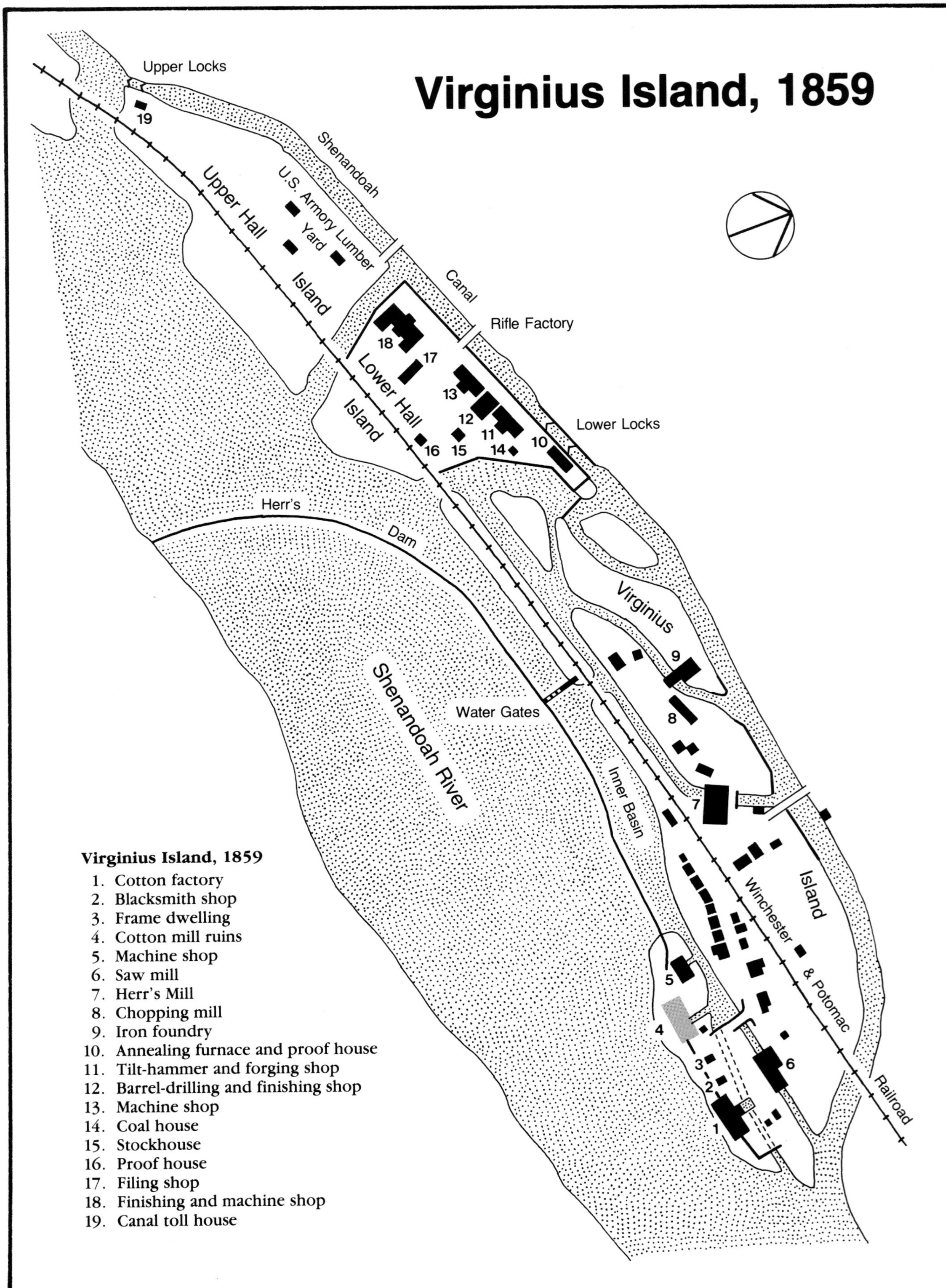
Virginius Island, 1859
Upper Locks
19
Shenandoah
Canal
U.S. Armory Lumber Yard
Upper Hall Island
Rifle Factory
Lower Hall Island
18
17
13
12
11
10
14
15
16
Lower Locks
Herr's Dam
Virginius Island
9
8
7
Water Gates
Shenandoah River
Inner Basin
Winchester & Potomac Railroad
5
4
3
2
1
6
Virginius Island, 1859
1. Cotton factory
2. Blacksmith shop
3. Frame dwelling
4. Cotton mill ruins
5. Machine shop
6. Saw mill
7. Herr's Mill
8. Chopping mill
9. Iron foundry
10. Annealing furnace and proof house
11. Tilt-hammer and forging shop
12. Barrel-drilling and finishing shop
13. Machine shop
14. Coal house
15. Stockhouse
16. Proof house
17. Filing shop
18. Finishing and machine shop
19. Canal toll house

Chapter Five

DEVELOPMENT OF TRANSPORTATION

In January 1837, the first trains passed over the Baltimore & Ohio Railroad's new bridge at Harpers Ferry. This combination railroad and wagon bridge had six timber spans which rested on five masonry piers. From abutment to abutment, the new bridge extended 830 feet across the Potomac River. Some records refer to this structure as the Wernwag Bridge after Lewis Wernwag, the builder who erected the bridge's timber spans. Other records refer to the bridge simply as "the new railroad viaduct."

The extension of railroad service to Harpers Ferry in 1836-37 facilitated an era of rapid growth and industrial expansion. On the "Ferry Lot Reservation" alone, The Gault House Saloon, the Wager House Hotel, the Virginia House, and a B&O ticket office were erected between 1837-43. The Harpers Ferry Armory and the several manufactories on Virginius Island could now obtain tools and heavy machinery promptly and cheaply from suppliers on the Eastern Seaboard. In turn, finished products could be shipped from these establishments east to Frederick and Baltimore or south on the new Winchester & Potomac Railroad to agricultural customers in the Shenandoah Valley. As the B&O Railroad extended its line westward, additional markets with rapidly expanding appetites for manufactured goods were opened up.

Indeed, the arrival of the railroad had a profound impact on the local economy at Harpers Ferry. But in fact, this development was just the last chapter in a gradual and deliberate evolution of transportation along the Potomac River.

In 1748, a group of tidewater landowners living along the Virginia banks of the Potomac estuary organized the Ohio Company. The shareholders in this enterprise, who included Augustine Washington and his sons Lawrence and George, set out to promote the settlement of western lands as far off as the Ohio River, and to develop a profitable fur trade across the Allegheny Mountains. Toward this end, the company pioneered an extensive trade route. From Alexandria, this route travelled overland by wagon road to Great Falls, continued up the Potomac River to Will's Creek at the present site of Cumberland, and snaked across the Allegheny Mountains along Nemacolin's Path—a horse trail hacked out of the wilderness by fur trader Christopher Gist and Indian scout Nemacolin.

For several years fur was the Ohio Company's principal item of trade. Other goods carried eastward were wheat, corn, lumber, and livestock. In return, textiles, firearms, hardware, and other finished goods travelled westward. Unfortunately, the French and Indian War, which flared up sporadically between 1755-63, brought destruction to company property and ruin to its profitable trade. Although the Ohio Company endured until 1798, it never fully recovered from the war.

In the fall of 1784, George Washington actively promoted the establishment of an all-water trade route up the Potomac River and its north branch. With Washington's personal encouragement, bills for the incorporation of the Patowmack Company were introduced in Maryland and Virginia. Both bills passed, and on May 17, 1785, the Patowmack Company was formally organized. The company's charter specified that navigation was to be improved from tidewater to the highest practicable point on the north branch of the Potomac River. Furthermore, the various improvements performed on the river were to permit the passage of river craft carrying 50 barrels of flour (about five tons) and drawing one foot of water in the driest season. Finally, the Potomac River's major tributaries were to be rendered navigable wherever practical.

The Patowmack Canal, then, was a system of river improvements around otherwise impassable obstacles. Where modest rapids occurred, boulders were removed and a suitable channel was created. Where substantial obstacles barred the way—such as at Little Falls and Great Falls—by-pass canals were dug and locks were erected.

George Washington was elected the Patowmack Company's first president in June 1785; by August he was conducting periodic inspection tours of the several rapids between Great Falls and Harpers Ferry. Engineers, described by a contemporary source as "local talent of doubtful quality," were soon excavating the necessary by-pass channels. In 1792, the first three by-pass canals, none of which required lift locks, were completed. Two were in the Harpers Ferry area: one at House's Falls (The Needles) and one at Payne's Falls (Whitehorse). The third was at Seneca Falls, 38 miles downstream. In 1795, temporary wooden locks were completed around Little Falls. The canal around Great Falls, however, consumed most of the Patowmack Company's time and financial resources. Here the river drops 76 feet in just 1,200 yards. Beginning in 1786, laborers started excavating a by-pass channel using hand tools and black powder. In addition,

Remains of an old lock built by the Patowmack Company along the Shenandoah River in about 1808. This particular lock is located one and one-half miles upstream from Harpers Ferry.

five lift locks were required to overcome the substantial fall. After much time and near bankruptcy, the company finally completed the Great Falls Skirting Canal in 1802. Seventeen years after it had been granted its charter, the Patowmack Company had at last rendered the Potomac River navigable upstream to the Savage River—a total distance of 218 miles and 350 yards from tidewater.

John Edwards Caldwell, in *A Tour Through Part of Virginia in the Summer of 1808,* described the various river craft that travelled down the Potomac River. These "batteaux" were from 70-75 feet long, four to five feet wide, and carried from 70-120 barrels of flour. The cost of shipping a single 196-pound barrel of flour was $1. Caldwell mentioned that accidents were rare, attesting to the skills of the boatmen. Travel, however, was usually limited to spring and fall, when water levels were high enough to convey these boats downstream. Other cargo not mentioned by Caldwell included furs, grain, corn, meat, lumber, pig iron, firearms, and less often, mountain whiskey. In the early 1800s, coal from western Maryland also became common river freight. Still, agricultural products comprised the principal cargo carried down the Potomac River.

With the completion of the Great Falls Skirting Canal, the Patowmack Company turned its attention to improving navigation on the Potomac River's principal tributaries. In 1806-07, engineers excavated a 580-yard canal behind Virginius Island on the Shenandoah River. This by-pass canal had an upper lift lock and a lower lift lock with a combined fall of 17 feet. By improving existing side channels of the Shenandoah River immediately above and below these two locks, a waterway 9,200 feet long was created. The Patowmack Canal, as this waterway was known locally, enabled boats coming down the Shenandoah River to by-pass Sawmill Falls (The Staircase). With the completion of four more by-pass canals and three other lift locks above Harpers Ferry, the Patowmack Company rendered the Shenandoah River navigable upstream to Port Republic—a total distance of 160 miles.

"Gundalows," as river craft in the Shenandoah Valley were known, commonly carried iron, flour, and lumber downstream to Harpers Ferry. Occasionally, these boats even proceeded to George-

Lock 33, yesterday and today. The top photograph was taken in 1876, and shows a loaded canal barge being towed downstream. The concentration of buildings indicates a great deal of activity here during canal days. The second photograph, taken in 1979, illustrates the toll time and nature have exacted on the old canal.
Top photo; National Park Service

town. Less frequent cargo included apples, potatoes, brandy, and corn from valley farms. During periods of high water, a "gundalow" would carry as much as 12 tons of iron or about 120 barrels of flour. For the four to five day trip down the river from Port Republic to Harpers Ferry, a boatman earned $14-$18. In the process he helped a crew of two to four men pole the loaded "gundalow" about 35 miles a day, and exercised remarkable skill in avoiding rocks and negotiating strong currents. Boats with cargo bound for Harpers Ferry often unloaded right at the foot of Union Street, just above the locks of the Patowmack Canal. In this manner a boatman avoided an additional toll. "Gundalows" then were usually broken up and sold for lumber—a return trip upstream being impractical.

In 1824, the Shenandoah Company took over the Patowmack Canal above Harpers Ferry. These new owners, in order to raise the water level in this canal, built a rubble diversion dam across the Shenandoah River at the upstream end of the waterway. As well, they replaced the canal's original two locks with a single lock of five feet lift and a double lock of 15 feet lift. In time, this waterway became known as the Shenandoah Canal.

The Patowmack Company, despite repeated appropriations from the legislatures of Maryland and Virginia, in fact never fulfilled the original terms of its charter—that river improvements permit the passage of boats carrying 50 barrels of flour in the driest season. Natural circumstance was the principal culprit; prolonged periods of low water and the annual onset of winter limited river commerce to no more than 45 days a year. By 1822, the Patowmack Company had spent $729,000, had placed itself into serious debt, and yet still had not met the minimum terms of its charter. Finally, a joint commission representing both Maryland and Virginia counseled against further state aid and instead recommended "a more effectual means of improving the navigation of the Potomac."

The financial deterioration of the Patowmack Company coincided with the birth of the canal era in America. From 1817-25, the Erie Canal was built in upstate New York. A year later the Pennsylvania Mainline canal system was begun. On July 4, 1828, construction commenced on the Chesapeake & Ohio Canal when President John Quincy Adams ceremoniously turned the first spadeful of dirt near Little Falls on the Potomac River.

What differentiated these canals from the Patowmack Canal was their manner of construction. Whereas the Patowmack Company had cleared channels and built by-pass canals around particular obstructions to river navigation, these new canals were level, reliable, artificial waterways. On adjacent towpaths mules pulled barges either upstream or downstream; feeder dams at appropriate intervals maintained a constant water level in each canal; lift locks raised or lowered barges travelling along these waterways as the adjacent river gradients required.

On Aug. 15, 1828, the Patowmack Company formally deeded its property and right-of-way to the C&O Canal Company. By November 1833, the C&O Canal was opened as far as Harpers Ferry. Here a river lock opposite the Shenandoah River admitted boats taking on cargo at Virginius Island or conveying agricultural products through the Shenandoah Canal. A large basin just above Lock 33 also served as a loading area for canal barges. Further up the canal, just above Dam No. 3, an inlet lock which admitted water into the canal also permitted barges with supplies for the Harpers Ferry Armory to cross the Potomac River and continue down the Armory Canal. In this manner, pig iron from Antietam Iron Works, for example, was transported to the armory workshops.

After numerous construction delays, labor difficulties, and cost overruns, the C&O Canal was finally completed in 1850. The total cost of the project was about $11 million. The finished canal extended 184 and one-half miles between Georgetown and Cumberland along the north bank of the Potomac River; plans to extend the waterway across the Allegheny Mountains to the Ohio River never advanced past the drawing board. Seventy-four locks, each with a lift of about eight feet, raised the canal a total of 605 feet on its course westward. Seven stone aqueducts carried canal and towpath across larger tributary streams. Paw Paw Tunnel, carved 3,118 feet through folded sedimentary rock, was the Canal Company's most ambitious enterprise. Here canal and towpath travelled one mile where the Potomac, in gargantuan loops, travels six.

Principal freight carried on the C&O Canal was coal, corn, wheat, and flour. These cargos were transported on mule-drawn barges 92 feet long and 14 and one-half feet wide. Each barge was capable of holding 120 tons—10 times the maximum load carried by a single flatboat using the old Patowmack Canal. Indeed, shipping freight on the C&O Canal was the cheapest method of transport up to that time. A barrel of flour, which in 1808 had cost $1 to ship down the Potomac River by "batteaux," cost as little as seven cents on the C&O Canal in 1831. Already, the economics of commerce were beginning to stimulate agriculture and industry along the Potomac River.

While canal construction in America was booming, there was a rush as well to build new roads and turnpikes. These roadways were stimulated by population expansion and by a growing need for

Lock 33 as it appeared shortly after the Civil War. Iron spans designed by Wendell Bollman now replaced the B&O Railroad's original timber bridge, destroyed by Confederates in 1861. Canal barges were loaded at the wide basin in the foreground. Harpers Ferry Historical Association

commerce between settlements not connected by river or canal. Turnpikes, or roads upon which a toll was charged, were eagerly built by stock companies organized for the purpose of making profit. The Lancaster Pike, which extended 66 miles between Philadelphia and Lancaster, was America's first turnpike. This roadway was built between 1792-94 at a cost of $465,000. The success of this well-travelled turnpike prompted the construction of toll roads and toll bridges alike. In 1806, the federal government became involved in internal improvements when Congress passed the National Road Act. Five years later, in 1811, construction commenced on the National Road, the first federally-financed turnpike in America. The first section of this road, which extended 131 miles across the Allegheny Mountains between Cumberland and Wheeling, was completed in 1818. Much of this turnpike followed the original path blazed by Christopher Gist and Nemacolin for the Ohio Company in 1752. The National Road, however, cost quite a bit more: about $13,000 per mile.

The turnpike boom soon came to Harpers Ferry. As mentioned previously, in 1824 Lewis Wernwag began erecting a "handsome double wooden highway bridge, 750 feet from one abutment to the other," across the Potomac River. This covered bridge was owned by the Wager family and appropriately was called Wager's Bridge. The Wager family, which was heir to the Potomac Ferry rights accorded to Robert Harper by the Virginia General Assembly, collected a toll from all foot and wagon traffic using its new bridge.

In 1830, the Harpers Ferry, Charles Town & Smithfield Turnpike Company was organized. A year later this turnpike was completed. Toll charged on this roadway was about two cents per mile for a horse and buggy, and was collected just outside of Harpers Ferry at a one-story frame toll-house. The turnpike entered town along Shenandoah Street and ended at the "Ferry Lot Reservation" near Wager's Bridge. Also in 1830, the Frederick & Harpers Ferry Turnpike Company was organized. By 1832, this toll road was completed from Frederick to the Maryland abutment of Wager's Bridge. A third turnpike company, the Hillsborough & Harpers Ferry Turnpike Company, was organized in 1851. Their toll road, which extended about 15 miles south into Loudoun County, Va., was completed in 1852.

In 1836, just a few yards downstream from Wager's Bridge, the B&O Railroad began erecting its combination railroad and wagon bridge. The following year, by an act of the Virginia Legislature all foot and wagon traffic was transferred from Wager's Bridge to the B&O bridge's wagon track. Nevertheless, the Wager family still owned the toll rights and continued to collect toll from all foot and wagon traffic using the new bridge. Finally, in 1839, the Wager family "conveyed all rights to take toll, including the rights incident to the original Ferry," to the B&O Railroad. By this same agreement, Wager's Bridge was dismantled.

In 1843-44, the Shenandoah Bridge Company erected a covered toll bridge across the Shenandoah River at the site of the "Shenandoah Rope Ferry Landing." This double-tracked wagon bridge had two stone abutments, two stone piers, and a wooden superstructure 375 feet long. Weatherboarding and cypress shingles protected the bridge from the elements. The federal government, which in 1818 had purchased the ferry rights here from the Fairfax family, for 25 years leased the operation of the rope ferry. After the new toll bridge was completed, the government still owned these ferry rights (they were not sold until after the Civil War). Consequently, the Shenandoah Bridge Company struck a deal: the government, in return for not exercising its ferry rights, was granted toll-free passage for all armory personnel and property. This arrangement was of some consequence, since the Harpers Ferry Armory leased the timber rights to 1,395 acres on Loudoun Heights and produced most of its charcoal there.

Railroad and canal were bitter rivals until the B&O became the Canal Company's major bond-holder after the Flood of 1889. Frequent floods, winter ice, and slowness of speed hampered the canal from the beginning. This photograph was taken near Miller's Bend, about two miles downstream from Harpers Ferry.

National Park Service

The turnpikes and toll bridges at Harpers Ferry mainly served the needs of local commerce: conveying farm goods and other raw materials into town; transporting agricultural tools, flour, and other finished products to nearby customers. An advertisement placed by the "Island Mill" in the *Virginia Free Press* on Aug. 18, 1831, drew particular attention to the Harpers Ferry, Charles Town & Smithfield Turnpike;

> *We will however, remark, that the fine road from Charles Town to Harpers Ferry, passing directly by this mill, is now very nearly completed, by which many of the farmers, particularly those quite near thereto, will be enabled for several months in the year to carry from 100 to 125 bushels of wheat at a single load—this to many will be a great advantage.*

Wagons for transporting cargo by road were drawn by as many as six horses and could haul up to 12,000 pounds—about half the load a "batteaux" could carry, and only one-twentieth the load of a single canal barge. Indeed, in 1822 it was eight times more expensive to transport a barrel of flour by wagon than to ship it by flatboat on the Potomac River.

The completion of the C&O Canal opposite Harpers Ferry in November 1833 was followed a year later by the arrival of the Baltimore & Ohio Railroad. Construction had begun on the B&O on July 4, 1828—the very same day that President Adams had turned a spadeful of earth to mark the start of the C&O Canal. Baltimore had invited its own hero to break ground for the railroad: Charles Carroll of Carrollton, the last surviving signer of the Declaration of Independence. Indeed, both railroad and canal would prove to be bitter rivals in the coming years.

At Point of Rocks, about 10 miles downstream from Harpers Ferry, the two companies engaged in their first dispute. Here the B&O Railroad secured an injuction against canal construction, contesting the right-of-way along a narrow shelf of land bordered by high cliffs and the Potomac River. In 1832, the courts ruled in the Canal Company's favor, based on the right-of-way which had been deeded over from the old Patowmack Company. But lacking much needed financial support, the Canal Company made a deal with the railroad whereby the B&O was permitted to proceed west to Harpers Ferry alongside the canal. The B&O, in turn, would subscribe to 2,500 shares of canal stock. It was further agreed that a tight board fence be erected between the canal and railroad to keep train engines from frightening canal mules. Rather than carry out this provision, however, the B&O instead compromised by limiting operations to horse power between Point of Rocks and Harpers Ferry until 1836.

On Dec. 1, 1834, the B&O Railroad reached the Maryland shore opposite Harpers Ferry. Meanwhile, in the Shenandoah Valley, construction proceeded on the Winchester & Potomac Railroad. This enterprise, which had begun laying track between Winchester and Harpers Ferry in 1833, commenced regular operation on March 14, 1836. The line extended a total of 32 miles up the Shenandoah Valley.

When the B&O Railroad completed its bridge across the Potomac River in January 1837, the tracks of both railroads connected in a straight line. In fact, the B&O had hoped to share a right-of-way with the Winchester & Potomac for a few miles up the Shenandoah Valley before turning west to Martinsburg. When negotiations toward this end failed, the B&O acquired a 10-foot right-of-way from the government through the armory grounds up along the Potomac River. In order to make the 90 degree turn from its straight bridge onto the masonry riverwall, which extended up the Potomac, a curved span or "Y" was added to the Harpers Ferry end of "the new railroad viaduct" between 1841-42. At about the same time, the bridge's timber spans were weatherboarded and covered with a tin roof.

Another conflict between the C&O Canal and B&O Railroad erupted soon after "the new railroad viaduct" was built. Apparently an agreement had been made between both parties whereby a tow walk would be erected along the downstream side of the bridge. This would permit mules to tow canal barges across the river to and from the Shenandoah River Lock. The tow walk was indeed built, but in 1837 a warehouse erected by the Wager family on the "Ferry Lot Reservation" blocked access to this walkway. The Canal Company tried to condemn the right-of-way through the Wager property, but without success. In 1841, the tow walk was dismantled; subsequently, the B&O charged the Canal Company 50 cents a ton for goods transported by rail across its bridge to the canal. Eventually, freight was carried to the canal by wagon, or barges were simply poled across the Potomac River to and from the Shenandoah River Lock when water levels permitted.

While the Canal Company struggled with construction difficulties and mounting debts, caused principally by the excavation of Paw Paw Tunnel, the B&O Railroad drove its tracks westward. In 1842, the railroad was opened to Cumberland—eight years before the canal was completed. In 1853, the B&O reached the Ohio River at Wheeling.

Commerce now was not only expanded to more distant markets, but shipping costs fell also as railroad and canal competed for business. Harpers Ferry indeed began to boom in the 1840s.

An 1859 photo showing the B&O's original railroad and wagon bridge at Harpers Ferry. The lines of the B&O and the Winchester & Potomac met inside the enlarged "Y" section of this bridge. Just beyond the bridge along the C&O Canal is a mule crossover bridge above the Shenandoah River Lock. National Park Service

This photo shows the old piers of the B&O's original timber bridge and subsequent iron Bollman Bridge.

Chapter Six

FOUNDATIONS OF THE FACTORY SYSTEM

The establishment of railroad service at Harpers Ferry created new opportunities for local industry. The U.S. Armory and the smaller manufactories on Virginius Island could now obtain raw materials and heavy machinery from distant suppliers at very reasonable cost. In turn, they could readily export their finished products to expanding domestic markets. But with opportunity there was also challenge—industry here had to remain competitive with other factories served by railroad, or lose customers.

To meet accelerating national demand for firearms, regional demands for other finished goods, and at the same time remain competitive, individual establishments introduced improved manufacturing methods. Productivity and economy were obtained by increased mechanization, a more efficient division of labor, and greater investments of capital. Hall's Rifle Works, established at Harpers Ferry before the railroad even arrived, was a harbinger of this transformation from craft to factory system, with results of national consequence.

On March 19, 1819, John Harris Hall of Portland, Maine, signed a contract with the War Department for the production of 1,000 breech-loading rifles. The contract provided Hall with a salary of $60 per month, established a royalty of $1 for each weapon he produced, and assigned him the use of shop facilities at the Harpers Ferry Armory. Hall commenced work in April 1819 at the Musket Factory, as the government workshops along the Potomac shoreline were subsequently called. Less than a year later, in February 1820, Hall moved his operations about a half-mile up the Shenandoah River into a small frame building adjacent to the lower locks of the Patowmack Canal. This structure, which had been used as a saw mill since the 1790s, became known as Hall's Rifle Works.

Hall's breech-loading rifle, designed and patented in 1811, had undergone rigorous testing by the War Department prior to the 1819 contract. Col. George Bomford, chief of ordnance, had "reported in favor of their adoption" in 1814. Three years later an ordnance officer in Massachusetts, after testing 100 of these patent rifles, had also "reported upon them in very favorable terms." Hall's rifle, it was found, took one-third the time to load as a conventional muzzle-loader, was lighter, and had greater accuracy with less recoil.

Hall spent much of his time in the early 1820s perfecting precision machinery for the manufacture of his rifles with interchangeable parts. Indeed, Eli Whitney of New Haven, Conn., and Simeon North of Middletown, Conn., had introduced some degree of interchangeability in arms manufacture in the early 1800s. But both men had met with only limited success. Hall, therefore, spent an equal amount of time constructing his machinery as he did making patent rifles. Not until the end of 1824 was his first parcel of 1,000 rifles completed. Nevertheless, the War Department was sufficiently impressed with the interchangeable component's of Hall's weapons that they ordered another 1,000.

Hall's unique breech-loading rifle was a considerable contribution to America's inventory of firearms. But his most enduring accomplishment was the successful introduction of the mechanized production of interchangeable rifle parts. Using precision machinery for stocking, forging, and cutting components, and an absolute minimum of hand work, Hall manufactured the first fully interchangeable weapons in America. This accomplishment laid a firm foundation for assembly line production—a principal ingredient of this country's developing factory system.

An investigative committee authorized by the secretary of war to look into every aspect of Hall's operation at Harpers Ferry, reported in January 1827 that

> *the machines, after the work is put into them, go thro' with the operation without any further* aid *from the boy, and when the operation is completed, give notice to the boy, who had been employed during the operation, in putting in and taking out work from other machines.*

More accolades for Hall's accomplishments followed. In 1832, Col. George Tolcott of the Ordnance Department wrote:

> *This manufactory has been carried to a greater degree of perfection, as regards the quality of work and uniformity of parts than is to be found anywhere—almost everything is performed by machinery, leaving very little dependent on manual labor.*

Hall indeed succeeded where no one else had. Self-acting machinery and interchangeable parts transformed the methods of arms manufacture from craft to factory system. A single workman, or even a boy with very little training, could, in Hall's own words, "perform more work than ten men with files, in the same time, and with greater accuracy." This system, carried to its logical conclusion,

1863-1864 photograph of the Musket Factory. The ravages of the Civil War are readily apparent from the workshops damaged or destroyed. The three main buildings that remain standing are, from left to right, the annealing shop and brass foundry, the smith and forging shop, and the boring mill. The Armory Canal is visible in the foreground. National Park Service

promised increased output and economy of labor, resulting in competitive prices for finished goods. While the initial capital investment of installing new machinery would be high, a mill owner adopting Hall's methods could expand production dramatically and recoup his costs in a short time.

Unfortunately, because the War Department was principally interested in Hall's mechanical ideas and production methods, only a small number of patent rifles were ever manufactured. As a result, the cost of producing each weapon remained high, and economy of production was never fully realized. Altogether, between 1823-1840, almost 20,000 patent rifles were assembled at the Rifle Works—about one-tenth the number of muskets produced at the Musket Factory during the same period.

On February 25, 1841, after a long illness, John Harris Hall died at the age of 60. In the same year, the War Department discontinued production of Hall's rifle, owing to objections raised by field commanders about its mechanical complexity. Nevertheless, Hall had made an enduring, if unceremonious, contribution to the development of America's factory system.

In March 1845, the government began extensive renovations of the workshops at the Harpers Ferry Armory. At the Musket Factory, a two-story boring mill, a forging and smith's shop, a stocking and machine shop (also called the "bell" shop), a polishing shop, a tilt hammer and barrel welding shop, a grinding mill, saw mill and carpenter's shop, and an annealing shop and brass foundry were erected between 1845-1851. In 1854, the government also built its own rolling mill. These several new workshops were constructed with durable cast-iron framing and brick superstructures, and were arranged to facilitate the movement of work from one stage of production to another (see map, page 33). New machinery was also installed for interchangeable manufacture.

Increased mechanization necessitated the use of more power. Consequently, the Armory Canal was enlarged and new millwork was installed. While at least one conventional overshot water wheel was still used to drive the machinery in both the "bell" shop and the boring mill, more efficient Boyden turbines (see Chapter 7) were installed in other buildings. At the grinding mill, saw mill and carpenter's shop, for instance, armory superintendent John Symington reported in 1851: "The turbine wheel, now in operation, will suffice for the whole machinery of this shop." Two more "iron turbine wheels" were also installed in the tilt hammer and barrel welding shop.

In about 1844, the government began tearing down the old frame shops of Hall's Rifle Works and replacing them with a new Rifle Factory. Between 1847-1852, a finishing shop, tilt hammer shop, and a large machine shop were all erected here (see map, page 34).

In overall physical appearance and day-to-day operation, the workshops of the Harpers Ferry Armory now assumed the proportions of a substantial arms manufactory. New machinery largely replaced the dexterous use of hand tools, and craft talent gradually disappeared. Production methods and work practices were standardized to the point of inflexibility. A 10-hour work day, for instance, was strictly enforced, and the consumption of alcohol during working hours was forbidden. Indeed, while productivity and economy of labor were generally improved, the individual workman became an inconspicuous component of this new industrial process. For better or worse, the factory system had taken root in Harpers Ferry.

Elements of the factory system soon appeared on Virginius Island. On Jan. 15, 1846, the Harpers Ferry and Shenandoah Manufacturing Company was incorporated by the Virginia Legislature. About a year later, on Jan. 7, 1847, the *Virginia Free Press* announced that this company was erecting a four-story cotton factory, 45 feet by 100 feet, on Virginius Island. In August 1847, the company extended its holdings on the island with the acquisition of the saw mill and machine shop from the firm Wernwag & Schofield. The purchase of another lot later that month consolidated about half the property on Virginius Island in company hands. No sooner was the cotton factory completed in March 1848, than the Harpers Ferry and

White Hall Tavern on Potomac Street was a favorite drinking house for armory workmen between 1847 and the Civil War. The building was erected in 1838-39, and enlarged about 10 years later.

The old cotton factory in 1900. About 50 years earlier, the Virginia Free Press *had described it as a large "Brick Factory Building, of the most permanent character, 104 feet by 48 feet, four stories high, with tin roof." The building was finally toppled by the Flood of 1924.* National Park Service

Shenandoah Manufacturing Company began to build another cotton mill just a few yards upstream. This second factory commenced operations in July 1849.

The first factories in America had indeed been cotton mills. In 1789, Samuel Slater emigrated from England to Pawtucket, R.I., and began erecting a textile mill. Slater had served an apprenticeship in an English textile factory, where he became familiar with the machinery of Richard Arkwright. Arkwright's spinning frame, which was patented in Nottingham, England, in 1769, produced a sturdy commercial grade of cotton thread durable enough for *warp*—the lengthwise thread in cotton cloth. Arkwright's machine was soon adapted to water power, and became known as a "water frame."

Arkwright also improved or invented machinery for the successive stages of carding, drawing, and roving cotton fiber into thread, and then combined these operations with spinning under one roof—an industry first. This original process became known as the Arkwright System of Cotton Machinery. It was this system and the machinery comprising it that Samuel Slater introduced in his Pawtucket mill. In 1790, Slater manufactured the first machine-made cotton warp in America, an achievement that stimulated the early development of this country's entire factory system.

Another contribution to America's textile industry was made by Boston merchant Francis Cabot Lowell. Just as Samuel Slater had learned the textile trade in one of Arkwright's mills before emigrating to this country, so Lowell saw fit to borrow from the English. In 1810-1812, he made a careful study of cotton machinery during a trip to England. In 1814, he built the first successful power loom in America, and with the aid of another engineer installed it together with spinning machinery in his Boston Manufacturing Company in Waltham, Mass. For the first time anywhere, the processes of spinning and weaving were brought together in the same mill. The "Waltham system," as Lowell's proc-

Cotton fiber, which passed from one cylinder of the carding engine to another, was rolled into layers and then combed into a thick rope or sliver. This sliver was coiled by a workman and taken to the drawing frame.

ess became known, was an important watershed in America's factory system.

Rapid improvements in cotton machinery followed and, together with the extension of railroad service, stimulated the textile industry from New England to the Southern states. Between 1840-1860, the number of spindles—a measure of the manufacturing capacity of textile mills—more than doubled. In fact, the textile industry grew nearly twice as fact as the population of the whole United States.

In this atmosphere of opportunity, the Harpers Ferry and Shenandoah Manufacturing Company's first cotton factory began producing cotton cloth by March 1848. This building was constructed of brick, fitted with gas lights, heated by steam through pipes, and equipped with the latest in textile machinery.

Beginning in the factory's carding department, a cotton opener and cleaner removed dust and burrs from the raw cotton. Two mechanical pickers then separated seed from fiber. These machines were descendents of Eli Whitney's cotton gin, which was invented in 1793. Next, the cotton was passed through one of 18 carding engines. Here, wire teeth fitted on cylinders combed out the cotton fibers, rolled the cotton into several layers called the *lap*, and then combed this lap into a thick rope or *sliver*. Three drawing frames stretched and tightened this loose, irregular sliver into a continuous thread called a *roving*. Six double roller beam speeders further refined the roving and wound the cotton filament onto several rollers.

The factory's spinning department contained 18 spinning frames with a total of 2,376 spindles—132 spindles per frame. Census records indicate that this number was about average for textile mills in the middle Atlantic states. These particular spinning machines were Danforth Cap Frames, supplied to the cotton factory by inventor Charles Danforth of Paterson, N.J. (see diagrams, pages 52-53). From the several rollers inserted on top of each frame, the roving ran onto bobbins placed on the spindles. These spindles in turn were mounted on lifters, which were given a reciprocating up and down motion by two endless cords attached to pulleys. When the frame was started, the whole operation commenced at one time: the bobbins on the spindles, rotating and at the same time rising and falling on the lifters, pulled out the roving from the rollers and spun it into warp, or twist-yarn. A unique contrivance supplied the spindles with oil and kept them constantly lubricated.

The warp was then taken to the weaving department. Here there were 97 looms, "driven by two new iron Turbine Wheels," according to an article that appeared in the *Virginia Free Press* on Dec. 9, 1852. Each power loom had a flying shuttle which moved swiftly from side to side, carrying a *weft* (the horizontal thread) back and forth across the warp as the odd and even threads were alternately lifted. The finished product was a finely woven cotton cloth.

The Harpers Ferry and Shenandoah Manufacturing Company opened its second cotton mill in July 1849. On Aug. 2, the *Virginia Free Press* ran the following announcement:

> *This factory is capable of manufacturing 400 lbs. of Cotton yarn, 1100 lbs. of Batting, and 50 lbs. of Candle wick per day. The building is of brick, 4 stories high—57 by 48 feet in the clear. The Machinery consists of seven Carders, four Spinning frames (called the Danforth Cap Frame,) two Reels, and Warning Mill, &c. This establishment employs some of the best and most experienced hands now in this country—some of them from establishments in Manchester, England.*

By the 1840s, textile machinery had been refined to such a degree that workers did little more than monitor the successive stages of production and transfer cotton filament from one machine to another. At cotton mills of this period, a work week generally consisted of 70 hours—12 hours a day, Monday through Saturday. In the middle Atlantic states, weekly wages for men averaged under $5; for women, maybe half that amount. Fortunately, room and board for a week usually cost no more than $2.

The atmosphere of opportunity that had stimulated the textile industry in the 1840s began to subside by the end of the decade. New cotton mill construction had outstripped the available supply of raw cotton. These factories as well were producing more cloth than the public could consume. The price of raw cotton consequently soared, while over-production seriously deflated retail prices. Industry profits shrank, and mill owners caught in the middle went broke.

Riding on this nationwide wave of optimism, the Harpers Ferry and Shenandoah Manufacturing Company had indeed expanded too fast. Finding

Power loom weaving in the 19th century. Young girls or unmarried women who ordinarily operated these machines looked upon work in a textile mill as an opportunity to earn money and escape the drudgery of farm life, not as a misfortune.

itself in serious debt, the company put its two mills up for sale on April 8, 1852. Just 10 days later, on April 18, the Flood of 1852 destroyed the cotton mill. Financially ruined, all the company's property was disposed of at a court-ordered public auction. In July 1854, Abraham H. Herr, who owned the three-and-one-half-story stone flour mill commonly called Herr's Mill, acquired all the former holdings of the Harpers Ferry and Shenandoah Manufacturing Company. With additional purchases in 1855, Herr became the sole owner of Virginius Island. The original cotton factory, which Herr had obtained with all its machinery intact, continued to operate under his supervision from 1855-1861.

The last new establishment to be erected on Virginius Island prior to the Civil War was a carriage and wagon manufacturing shop, opened in 1857. "Rohr & Brother," who operated this shop, advertised in the *Virginia Free Press* on July 26, 1860, that:

> *...they have on hand, and are now furnishing an assortment of*
>
> *CARRIAGES, ROCKAWAYS AND BUGGIES,*
>
> *Which they will sell on reasonable terms. Their Carriages, &c. are made of the best materials, and finished in the best possible style, and warrented for twelve months.*
>
> *WAGONS of all sizes, from Six Horse down to the light Express Wagon, on hand or made to order, which they will insure for twelve months....*

This carriage and wagon manufacturing shop occupied a small brick building adjacent to Herr's Mill.

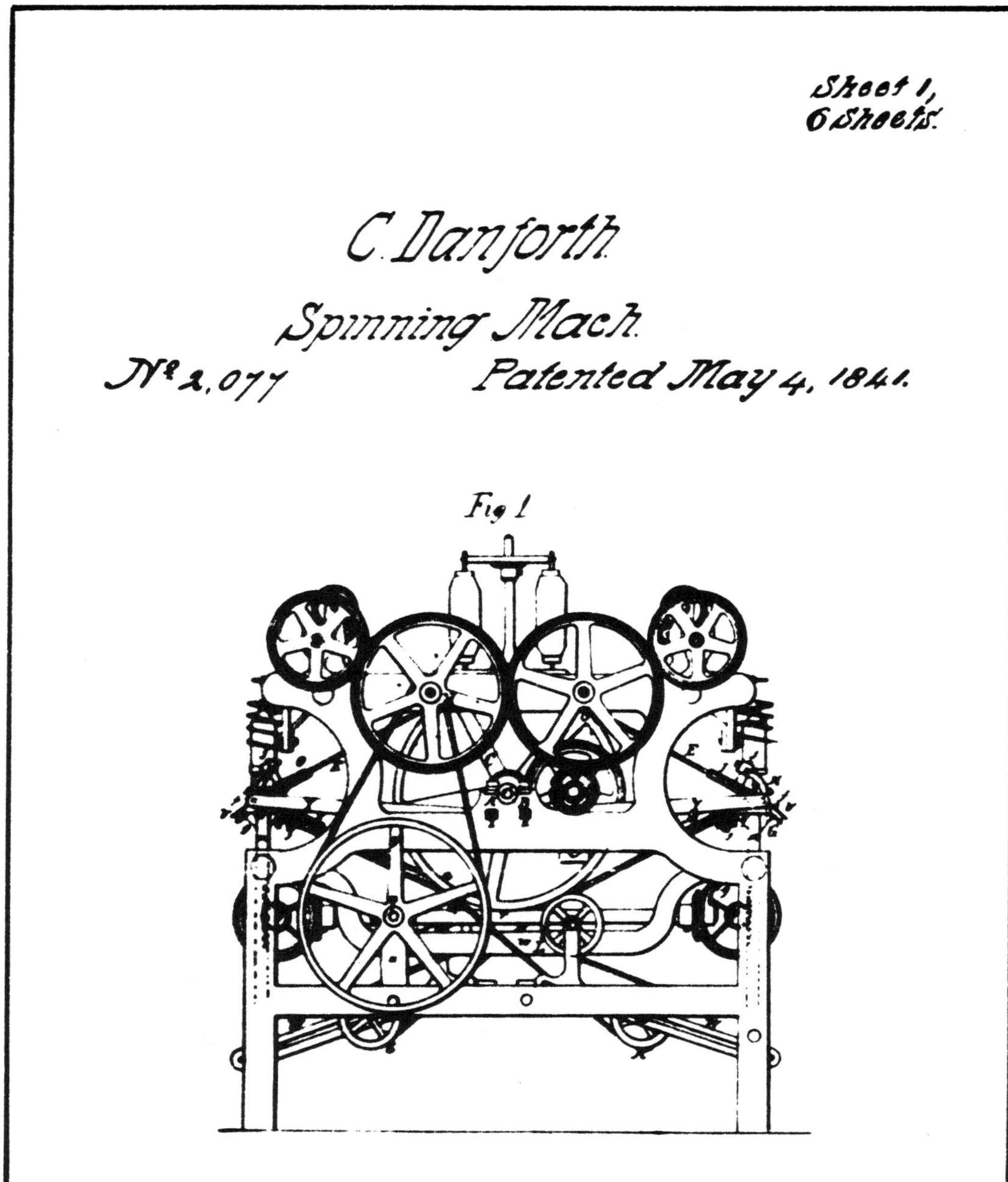

Sheet 2
6 Sheets.

C. Danforth.

Spinning Mach.

No 2,077. Patented May 4, 1841.

Fig 2

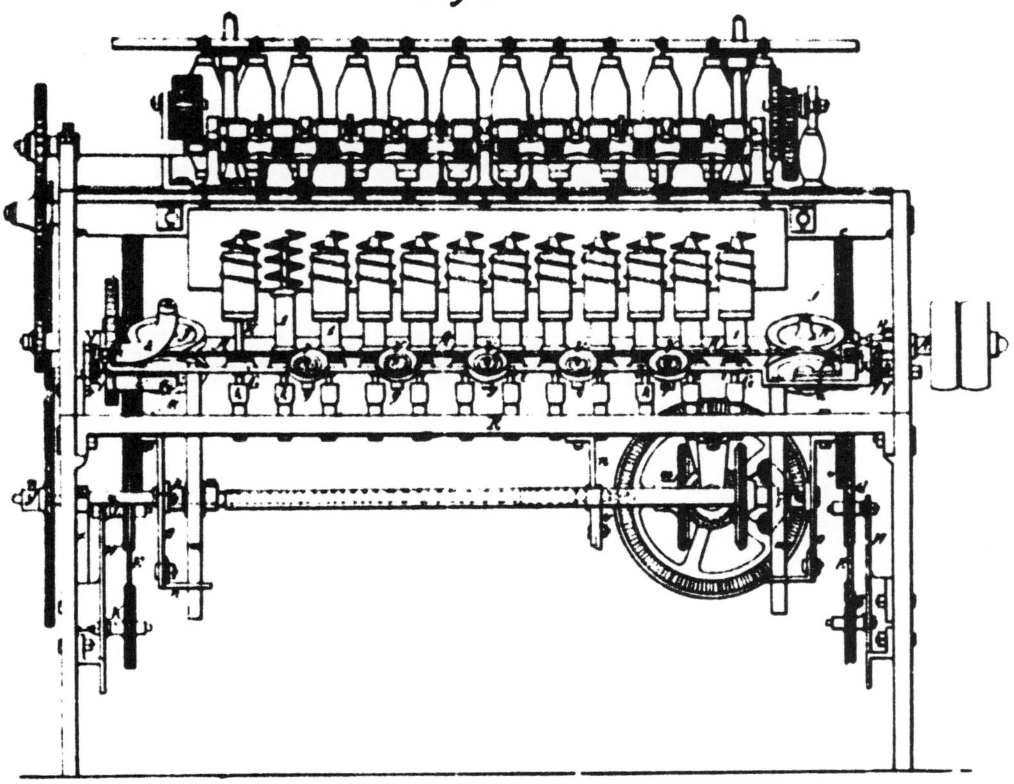

Profile (left) and front view (above) of the Danforth Cap Frame. This spinning machine, patented by Charles Danforth on May 4, 1841, produced warp or twist yarn used for the manufacture of cotton cloth. The rollers are located on top of the frame; the bobbins are affixed to the spindles below. United States Patent Office

These two underground tunnels once conveyed water to the cotton factory on Virginius Island.

Chapter Seven

NEW WATER WHEEL TECHNOLOGIES

The proliferation of new machinery for America's rapidly expanding factory system created demands for more power. Even overshot water wheels, which were able to convert about 60 percent of the energy available in falling water into motive power, were no longer adequate for the large new manufactories. At the same time, the deforestation of drainage basins caused by the expansion of industry and agriculture greatly aggravated seasonal fluctuations in streamflow. Mill operators consequently were forced to extract the utmost advantage from the little water they sometimes had.

Benoit Fourneyron of France is credited with inventing the first turbine in 1827. Water, which entered Fourneyron's unique circular device at the center, flowed outward through an inner ring of fixed guide vanes, and struck the moving vanes of an outer runner. The vanes of this outer runner were curved in the opposite direction from the inner guide vanes. This arrangement reversed the direction of water flow within the turbine, created a reactive force, and extracted more energy from the falling water. Indeed, Fourneyron's turbine was found to be as much as 80 percent efficient in converting the energy available in the water into power.

The advantages of Fourneyron's turbine over conventional water wheels were readily apparent to inventors and mill operators alike. Undershot, breast, pitch-back, and overshot wheels all generated power through the direct impact or weight of falling water on buckets or floats. In none of these arrangements did water move relative to the wheels. In a turbine, however, water moved through ducts and vanes and developed into velocity energy as impulses and reactions were set up between the flowing water and the turbine passages. Consequently, more energy could be extracted from the same "head" and "fall" of water.

As improvements were made in subsequent turbines, additional advantages over conventional water wheels were obtained. All-iron construction, for example, created a strong, durable and generally maintenance-free mechanism. As turbines became more compact, two or more of these devices could be used in place of just one water wheel, with a substantial advantage in generating power. Turbines also worked submerged, freeing them from the direct interference of winter ice, and enabling them to efficiently utilize considerable variations in head.

In the cotton factory on Virginius Island, "two new iron Turbine Wheels" of Fourneyron design were installed in about 1850. These particular turbines were manufactured by E.C. Kilburn & Company of Fall River, Mass., which had adopted the Fourneyron wheel from Europe in 1844. The turbines in the cotton factory were 5 feet, 10 inches in diameter, and were purchased for about $575 each.

At about the same time these two turbines were installed, the Harpers Ferry and Shenandoah Manufacturing Company erected a rubble dam across the Shenandoah River at Virginius Island. This new structure was built to furnish a more reliable supply of water to the company's cotton factory, cotton mill, saw mill, and machine shop. They also built an extensive river wall to form an inner basin, where water was temporarily stored before being distributed to individual headraces. Three brick-lined arches, each fitted with gates, regulated the amount of water supplied from the river into this inner basin. One of the most interesting features of these elaborate new waterworks was the headrace that led to the cotton factory. Here, water was conveyed from the inner basin through two underground tunnels. This arrangement helped maintain pressure on the water, and as a consequence, powered the factory's new turbines with greater effect.

The rubble dam across the Shenandoah River was rebuilt or repaired several times as natural circumstances required. It subsequently became known as Herr's Dam, after Abraham H. Herr, who by 1855 had acquired all the property on Virginius Island.

In 1845, Uriah A. Boyden of Lowell, Mass., greatly improved the Fourneyron turbine by adding a conical approach passage for the incoming water. In this manner, the head of water was put under pressure and admitted into the center of the turbine with a greater velocity. This arrangement in turn created a more forceful reactive impulse, and extracted more energy from the falling water. The advantages of this design led to the rapid replacement of conventional water wheels with Boyden turbines in Massachusetts and elsewhere.

In 1851, at least three Boyden turbines from the Ames Manufacturing Company of Springfield, Mass., were installed at the Harpers Ferry Armory. These new turbines replaced breast wheels, which had been used since the first forging shop was completed in 1800.

Even before these Boyden turbines were acquired, the government had expended much

Illustration of a Boyden Turbine, circa 1845. The introduction of Boyden's conical approach passage greatly improved the effectiveness of the outward flow runner vanes. Smithsonian Institution

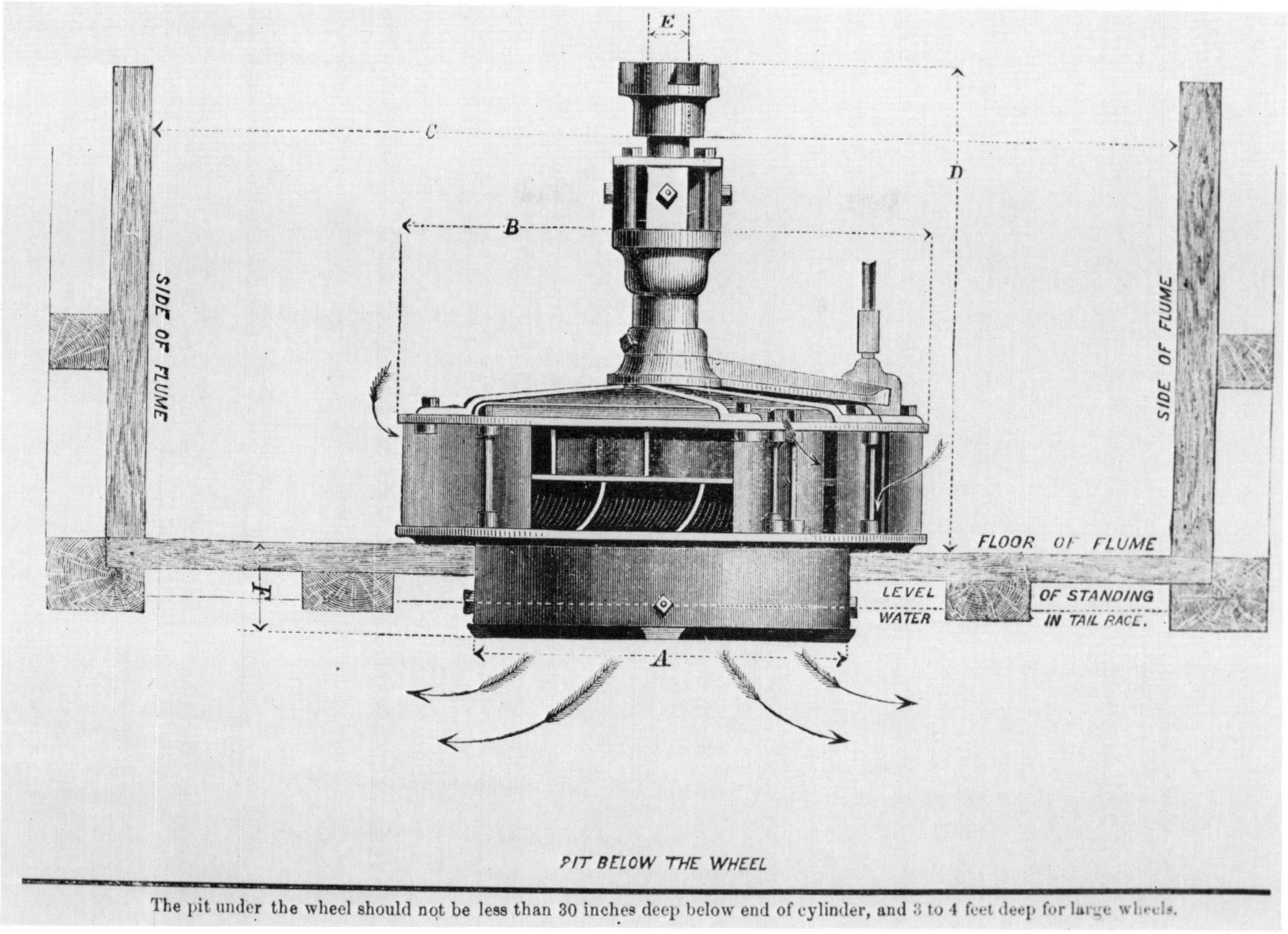

James Leffel's Double Turbine Water Wheel, from an 1870 catalog. By combining radial inflow vanes with axial discharge vanes, Leffel was able to extract more power from the falling water.

Ohio Historical Society

money and effort to improve the Armory Dam and Armory Canal, in order to obtain a more reliable supply of water to power heavy arms-making machinery. The Armory Dam, which had originally been built in 1799-1800, underwent extensive renovations during the years that followed. In 1828, for instance, Lewis Wernwag reconstructed the entire dam, using timber piers bolted together with iron spikes and filled with stone—a common arrangement for milldams in the 19th century. When the C&O Canal was being extended to Harpers Ferry in 1832-33, the dam was enlarged so that water could be provided both to the Musket Factory and the new canal. Because this was the third of seven dams along the Potomac River used to supply water to the canal, it was called Dam No. 3 in Canal Company records. To further improve the supply of water conveyed to the Musket Factory after the Boyden turbines were installed, the government spent $7,000 in 1856-57 for extensive dam repairs.

Still, the Armory Dam was apparently inadequate, for on June 1, 1859, the government began laying a masonry foundation for a new dam several yards downstream. By September 1860, about one-third of the proposed 1,470-foot-long structure was completed on the Maryland side of the Potomac River. In April 1861, construction was interrupted by the outbreak of the Civil War, however, and the new dam was never completed.

Between 1830-60, the government also spent approximately $68,000 on rebuilding and repairing the Armory Canal. A sample of the work performed during this period is contained in an 1844 government report:

> *During the past Season, the canal which supplies the power from the Dam on the Potomac, has received extensive and thorough repairs, in new Head Gates, a Pier head and Ice breakers, and new Gates to the Lock which admits boats with Coal and other supplies for the Armory; indeed, the whole mass of masonry at the head of the Canal, and lock, have been renewed and the Canal cleared out—whereby the power will be much increased in periods of low water, and a more uniform supply of water be obtained. . . .*

These iron driveshafts are connected to four Leffel turbines which still sit in the ruins of the Child & McCreight flour mill on Virginius Island. The turbines originally rested on a wooden floor, through which the water was discharged into the river below.

Both the Fourneyron turbine and Boyden turbine were of *outward flow* design—the water, being admitted at the center, flowed outward through the wheel. This arrangement, however, had a detrimental effect on the efficiency of the turbine. Because its passages diverged the flow of water became separated, resulting in a loss of energy due to eddy formation.

Consequently, *inward flow* turbines of the type first patented by Samuel B. Howd in 1838 became increasingly popular. In these turbines, water was admitted from the outside circumference of the wheel through a ring of fixed guide vanes. The water was then passed onto inner runner vanes before it exited through the bottom of the wheel. Because the pressure on the incoming water was maintained and even intensified as it passed through the wheel, a more effective reactive force was imparted. In about 1849, Massachusetts engineer James B. Francis designed a more durable and efficient inward flow turbine, which was used with success at a variety of mill-seats.

In 1862, James Leffel of Springfield, Ohio, departed from the common design of the Francis turbine. Leffel's Double Turbine Water Wheel instead used a double runner. The upper runner had radial inflow vanes of the Francis type, while the lower runner consisted of curved vanes which re-directed the falling water in an axial direction as it was discharged from the turbine. This arrangement resulted in a turbine with *mixed flow*. Leffel asserted in his patent application that his wheel was capable "of yielding from ninety-two to ninety-five per cent. of the power of the water and a greater per cent. than any other wheel heretofore constructed."

Another improvement which Leffel introduced was a series of gates around the circumference of the turbine which were opened or closed through linkages operated from above. This arrangement allowed a mill operator to shut off one or more turbines without having to shut down the entire water supply to other wheels operating at the same flume.

In 1867, when the firm of Child & McCreight began converting the old cotton factory into a flour mill (see Chapter 9), four Leffel turbines were installed here. These iron water wheels, which together produced about 300 net horsepower, drove 10 pair of burr-stones for more than 20 years.

A further improvement in turbine technology was achieved by John B. McCormick in about 1876. By adding spoon-shaped discharge sections to the bottom of a turbine's runner vanes, McCormick found that he could extract more energy from water leaving the wheel. Indeed, so effective was McCormick's design, most subsequent turbine manufacturers used these spoon-shaped discharge vanes on their own wheels.

One such company was the Globe Iron Works of Dayton, Ohio, which on Feb. 19, 1884, patented the New American Turbine Water Wheel (see diagrams, pages 60-61). The runner of this turbine, however, was comprised of a solid or continuous casting, without a rivet or bolt on any part of the wheel—a substantial improvement over previous constructions. In their catalog, the Globe Iron Works claimed that they had created "a perfect wheel, with even, true, and smooth surfaces, and of unquestioned strength, and which we will guarantee to stand the pressure of any head."

The wheel case of this turbine was also improved with the addition of graduated inflow chutes or gates. These hinged chutes, when opened or closed, admitted water onto the runner in gradually expanding or contracting veins. In this manner, machinery coupled to these turbines could be started up slowly as the mill operator gradually opened the chutes on the wheel case. This reduced the stress normally created on belts and shafts by rapid start-ups.

Another company which adopted McCormick's discharge vanes was the S. Morgan Smith Company of York, Pa., which manufactured the Improved Success Water Wheel. The runner of this turbine was comprised of long buckets which were gradually angled forward from top to bottom. This arrangement permitted the smooth passage of water through the wheel, without any abrupt bends or turns. The runner was also fitted up closely to the wheel case, so that the long, curved buckets formed perfect pockets and prevented any loss of power from water discharging prematurely.

When Thomas Savery of Wilmington, Del., erected both a pulp mill and a paper mill at Harpers Ferry between 1887-90 (see Chapter 9), he chose New American turbines and Improved Success turbines to power his machinery. A total of eight New American turbines and two Improved Success turbines in the pulp mill produced 1,600 net horsepower—an average of 160 horsepower per wheel.

Water wheel technology had indeed come a long way since the days of conventional undershot, overshot, and breast wheels. Improvements in turbine design and construction had effectively kept pace with the increased power needs of an expanding factory system. But there were still other factors to contend with.

(No Model.) 3 Sheets—Sheet 1.

W. M. MILLS.

TURBINE WATER WHEEL.

No. 293,904. Patented Feb. 19, 1884.

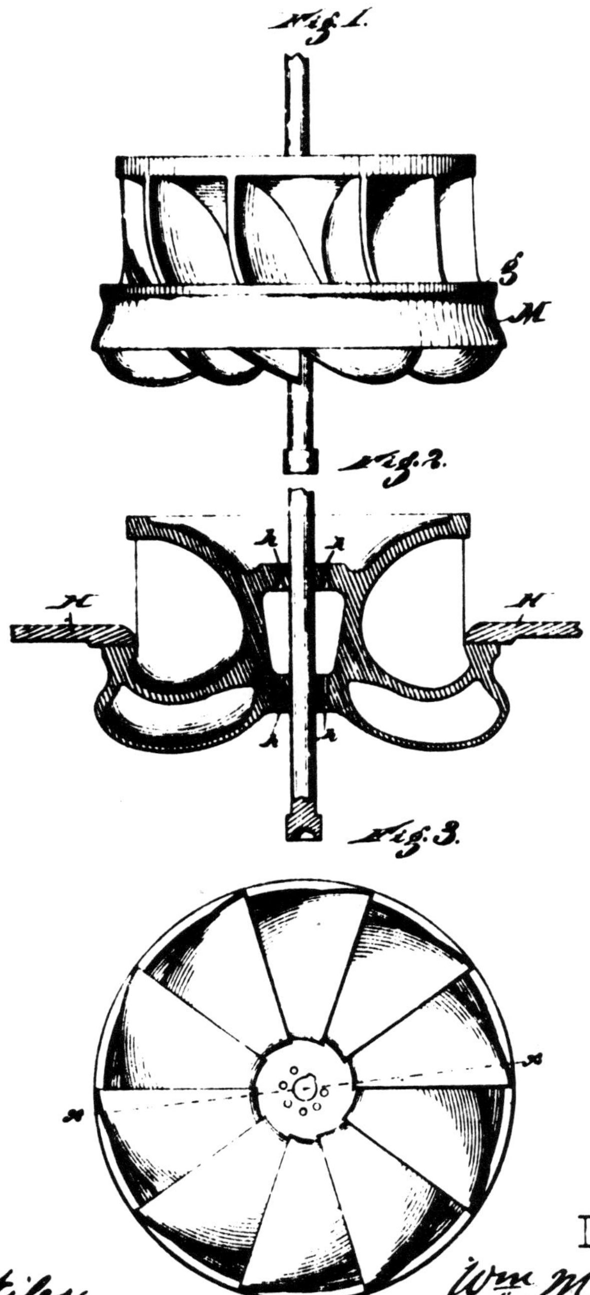

Attest,

Jno. B. Miles.

C. W. Rector

Inventor,

Wm. M. Mills

by Stevens & Co.

his Attys.

Runner of the New American turbine, manufactured by the Globe Iron Works of Dayton, Ohio. Note the spoon-shaped discharge vanes at the bottom of the wheel. United States Patent Office

(No Model.) 3 Sheets - Sheet 3.

W. M. MILLS.

TURBINE WATER WHEEL.

No. 293,904. Patented Feb. 19, 1884.

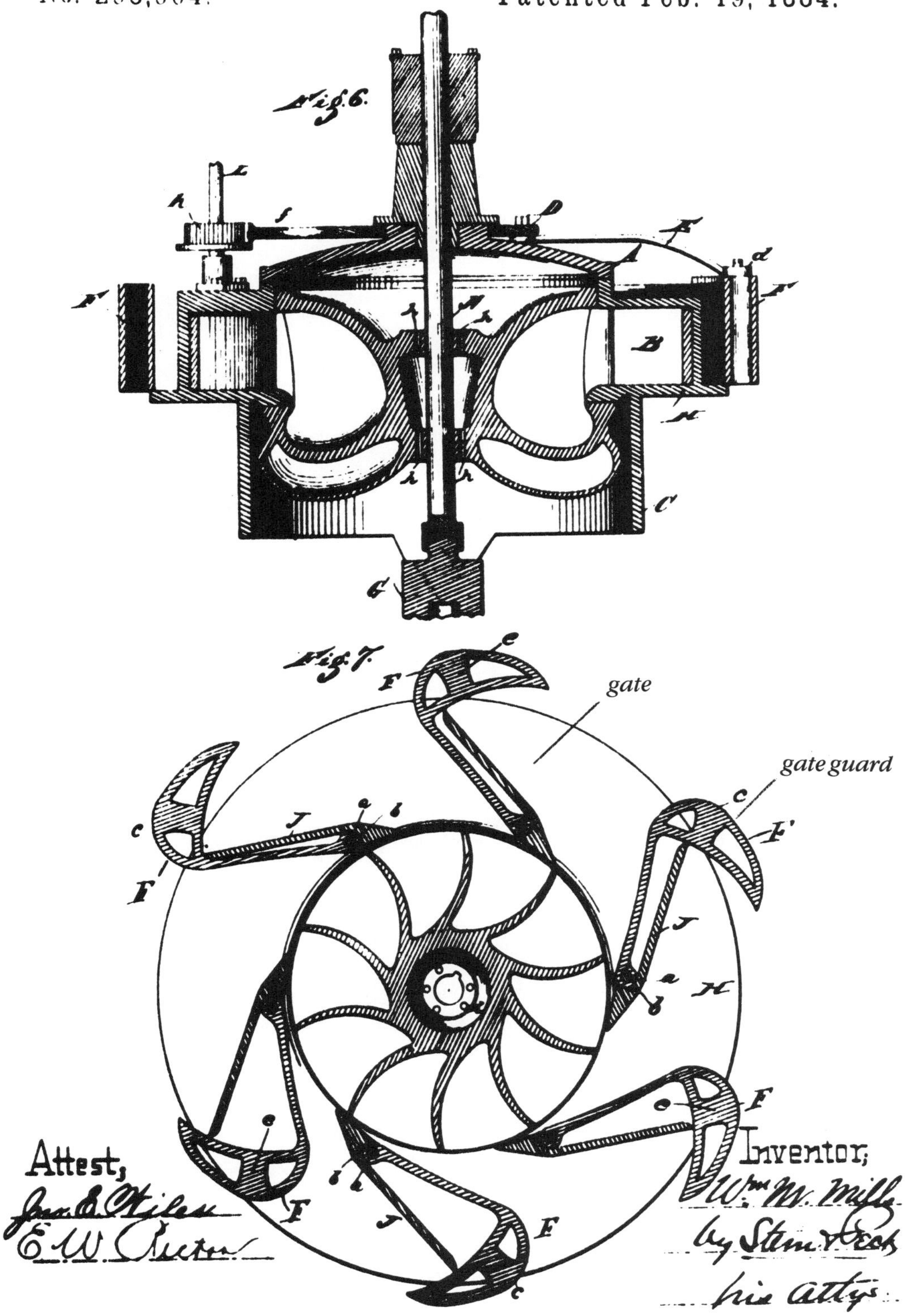

These cutaway diagrams illustrate the arrangement of graduated chutes or gates around the runner of the New American turbine. The fenders or gate guards attached to the outside of the gates relieved them from the pressure of the incoming water and permitted them to be opened or closed with relative ease.

United States Patent Office

John Brown's Fort in Harpers Ferry was originally built in 1848 as the Musket Factory's fire enginehouse and watchman's office. Here John Brown was cornered and captured by U.S. Marines on Oct. 18, 1859. A notable landmark and popular tourist attraction, this was the only armory building to escape destruction during the Civil War.

Chapter Eight

FORCES OF DESTRUCTION

Perhaps the first indication that Harpers Ferry would be in for hard times came on April 18, 1852. On that day, the Potomac and Shenandoah rivers rose to an unprecedented height, inflicting substantial damage on low-lying property. The *Virginia Free Press,* on April 22, described this flood in some detail:

> *Our country has been visited by an awful calamity.—The oldest inhabitants never before witnessed such a rise in the waters of the Potomac and Shenandoah river.—The inundation came upon us like an avalanche—sudden, unexpected and overwhelming.*
>
> *It commenced raining about 4 o'clock P.M. on Saturday last, and continued without intermission, until Tuesday at noon. It fell, however, so gently that no one anticipated more than an ordinary swelling of the streams. . . .*
>
> *By Tuesday morning the streets of Harpers Ferry, had become channels for mighty rivers, and the second and third stories and house-tops, resting places for its distressed inhabitants. . . .*
>
> *The destruction of property is beyond all description. Every house on Shenandoah and Potomac streets was almost entirely submerged—the water being six feet higher than at any other period within the recollection of man. . . .*

In addition to the extensive damage suffered by residential property, the wagon bridge to Virginius Island was swept away, some trestle work and roadbed on the Winchester & Potomac Railroad was washed out, and the cotton mill belonging to the Harpers Ferry and Shenandoah Manufacturing Company was destroyed. But as the floodwaters receded, townspeople repaired the damage and went about their daily lives, untroubled by political rumblings in other parts of the country.

On July 3, 1859, a certain Isaac Smith arrived at Harpers Ferry and, with his two sons, Owen and Oliver, rented a farm about five miles away in the Maryland countryside. Not even the smallest stir was created as several other men gathered here over the next few weeks. In fact, Isaac Smith was John Brown, and at the Kennedy Farm he and several followers were secretly planning a raid on the Harpers Ferry Armory. With the guns that were stored there, Brown intended to arm slaves and lead them to freedom. Three years earlier, his cold-blooded murder of five pro-slave Kansans had demonstrated his fanaticism on the issue.

Brown and 18 men of the "Provisional Army of the United States" made their move on Sunday evening, Oct. 16, 1859. Armed with Sharps rifles, the raiders crossed the B&O railroad and wagon bridge, captured the bridge watchman, and quickly seized the armory grounds. Men were then sent to seize the Rifle Factory and to watch the Shenandoah Toll Bridge. Despite Brown's efforts not to stir the townspeople, an alarm was soon sounded. At midnight a shot from a raider's gun narrowly missed the bridge watchman's relief man. Then at 1:30 a.m. the eastbound express train from Wheeling to Baltimore was detained. In the commotion that followed, Heyward Shepherd, the station baggageman and a black freedman, was shot and mortally wounded by one of Brown's men.

As word of the raid spread, militia companies from Maryland and Virginia moved into town during the daylight hours of Oct. 17. Several raiders were captured or killed. Brown, the remaining raiders, and a handful of hostages were then cornered in the Musket Factory's fire enginehouse. Late that evening, 90 United States Marines, summoned from Washington, D.C., joined in Sandy Hook by Col. Robert E. Lee and Lt. J.E.B. Stuart, entered the armory grounds and replaced the disorganized militia.

At daybreak on Oct. 18, Lt. Israel Green picked a storming party of 12 Marines. When Brown turned down a surrender demand, Green and his men battered down the door of the fire enginehouse. In about three minutes, the fighting was over. One Marine had been killed, one wounded. Brown, though injured, was taken alive. Of the 19 raiders who entered Harpers Ferry, 10 were killed, five were captured, and four escaped. In addition to Heyward Shepherd and the one Marine, four townspeople were also killed.

John Brown, charged for "conspiring with slaves to commit treason and murder," was convicted and hanged in nearby Charles Town on Dec. 2, 1859. During a pre-trial interrogation, Brown issued somber words which proved to be a prophetic warning: "You may dispose of me very easily . . . but this question is still to be settled—this negro question I mean—the end of that is not yet."

And so it wasn't. In November 1860, Abraham Lincoln was elected our nation's sixteenth president. Southern states, outraged by the prospect of a Republican chief executive, leaned toward secession, and on Dec. 20, South Carolina withdrew from the Union. In January 1861, Alfred M. Barbour, superintendent of the Harpers Ferry Armory, advised the War Department that he had "reason to apprehend that some assault will be

Ruins of the Musket Factory's smith and forging shop in about 1882. The B&O Railroad's new iron Bollman Bridge, designed and built by Wendell Bollman, replaced the old covered timber bridge, which the Confederates blew up in 1861.

National Park Service

made upon the United States Armory at Harper's Ferry." In fact, a company of United States regulars had been stationed here to protect government property since John Brown's raid.

The bombardment of Fort Sumter and Lincoln's

call for 75,000 volunteers to put down the rebellion finally prompted Virginia to act. On April 17, 1861, a state convention, after several months of indecision, passed an ordinance of secession. The Harpers Ferry Armory became an immediate military objective.

Lt. Roger Jones of the United States Army defended the armory with 50 regulars and 15 volunteers on the evening of April 18. In Charles Town, just eight miles away, several companies of Virginia militia—360 men in all—assembled and advanced toward Harpers Ferry. Jones, outnumbered and unable to obtain reinforcements, had torches set to kegs of powder in the Musket Factory and arsenal buildings. At about 10 p.m., an explosion ripped through the large arsenal. Townspeople acted quickly to extinguish fires in most of the Musket Factory workshops, but by the time the first Virginia militiamen entered town, the carpenter's shop, large arsenal, and some 15,000 arms were consumed by fire. The Civil War had indeed come quickly to Harpers Ferry.

A few days later, Thomas J. Jackson, with orders from the governor of Virginia, arrived at Harpers Ferry as a colonel of Virginia volunteers. Jackson immediately set his 8,000 motley soldiers to daily drill and target practice. He as well began the pressing task of disassembling the armory's heavy machinery and shipping it to Confederate installations in Richmond and in Fayetteville, N.C. Altogether, Jackson's men confiscated 4,287 firearms, plus enough components to assemble as many as 10,000 more, 3,000 machines, thousands of feet of belting and shafting, and some 57,000 assorted tools, all from the Musket Factory. Another 132 machines and countless tools and spare parts were also removed from the Rifle Factory. Even locomotives and rolling stock seized from the B&O Railroad were transported up the Shenandoah Valley.

Brig. Gen. Joseph E. Johnston replaced Jackson as commander at Harpers Ferry on May 24, 1861. By mid-June, the Confederates had packed up the last shipments of tools and arms-making machinery. At 4 a.m., June 14, 1861, Johnston's troops blew up the B&O railroad and wagon bridge, burned the principal Musket Factory workshops, and withdrew south toward Winchester. Two weeks later, on June 28, Confederate raiders burned the Rifle Factory and the Shenandoah Toll Bridge. In just two months of war, Harpers Ferry had lost its two largest factories and the services of its main transportation arteries.

Another establishment that soon met the torch was Herr's Mill on Virginius Island. In October 1861, Abraham H. Herr, a Union sympathizer, invited federal troops camped across the Potomac River to remove a substantial supply of unground grain from his flour mill. Otherwise, he feared it might fall into Confederate hands. The 13th Massachusetts and 3rd Wisconsin regiments obliged, and the grain was removed in a few days. Nearby Confederate forces, however, were informed of this action, and a few days later set fire to Herr's three-and-one-half-story mill. When the flames finally died out, only its four sturdy stone walls remained standing.

While the tide of war ebbed and flowed along the Potomac boundary separating Maryland and Virginia, Harpers Ferry was inevitably caught in the middle. The town guarded a key railroad bridge on one of the Union's principal east-west supply corridors. Indeed, as the different armies came and went, this bridge was destroyed and replaced nine times in the course of the war. Harpers Ferry also sat at the very foot of the Shenandoah Valley—the "breadbasket of the Confederacy." Union control of this strategic town exposed the soft underbelly of the whole valley. Conversely, Confederate occupation of Harpers Ferry threatened the flank and rear of Washington, D.C., just 60 miles to the east.

So the Union forces dug in, first on the Maryland side of the river, using 1448-foot Maryland Heights as their primary defensive position. On Feb. 7, 1862, federal troops burned the several buildings on the "Ferry Lot Reservation" at the confluence of the Potomac and Shenandoah rivers. Here, Confederate sharpshooters had routinely been taking cover while firing on Union positions. Later that same month, federal forces occupied the whole town and established additional defensive positions on Bolivar Heights. By September 1862, Col. Dixon S. Miles commanded a strong Union garrison of about 12,000 men at Harpers Ferry.

Meanwhile, on Sept. 4 and 5, Robert E. Lee's Army of Northern Virginia was splashing across the Potomac River about 20 miles downstream. The South's first invasion of the North had just begun. The garrison at Harpers Ferry, however, posed a substantial threat to the Confederate rear; Colonel Miles could harass or even sever Lee's supply lines from Virginia. Lee, therefore, decided to divide his army. The corps of James Longstreet and D.H. Hill were ordered from Frederick, Md., toward Hagerstown in preparation for an invasion of Pennsylvania. The now prominent Maj. Gen. Thomas "Stonewall" Jackson, who had spent a month at Harpers Ferry early in the war, was sent with 32,000 men to surround and capture this strategic town.

Jackson's forces advanced in three separate wings. One wing, under his direct command, marched from Frederick to Boonsboro, crossed the Potomac River at Williamsport, captured Martinsburg, and on Sept. 14, pitched camp just west of Bolivar Heights behind Harpers Ferry. A second wing, commanded by Brig. Gen. John G. Walker, crossed the Potomac River at Point of Rocks and proceeded to Loudoun Heights. Because these heights were well within range of federal batteries on Maryland Heights, Colonel Miles had neglected to post any men or artillery here. Walker therefore moved a battery of artillery up onto Loudoun Heights without opposition, and on Sept. 14, he exchanged the first artillery fire with Union guns.

The third wing in Jackson's advance was commanded by Maj. Gen. Lafayette McLaws. McLaws understood the topography around Harpers Ferry well. "So long as Maryland Heights was occupied by the enemy," he wrote, "Harper's Ferry could never be occupied by us. If we gained possession of the heights, the town was no longer tenable to them." Confederate infantry advanced south along the crest of Elk Ridge—the northern extension of Maryland Heights—and on September 13, drove some 2,000 Union defenders off the mountain. In this critical skirmish, for which a Union colonel was later cashiered for misconduct, the federal forces had lost their only advantage. Harpers Ferry was now completely surrounded.

Confederate artillery fire upon Harpers Ferry was effective and demoralizing. Col. William H. Trimble of the 60th Ohio wrote that there was "not a place where you could lay the palm of your hand and say it was safe." The Union garrison, 12,693 men strong, finally surrendered on the morning of Sept. 15. Some 13,000 arms and 73 pieces of artillery were also seized.

When federal forces reoccupied Harpers Ferry several days later after the bloody Battle of Antietam, they better appreciated the importance of the high ground around the town. Engineers proceeded to erect strong fortifications and gun emplacements on Maryland Heights, Loudoun Heights, and Bolivar Heights. Except for minor skirmishing, and an unsuccessful bid by Lt. Gen. Jubal A. Early to dislodge Union defenders from Maryland Heights in July 1864, Harpers Ferry was spared further fighting for the duration of the war.

Still, the Civil War had already exacted a heavy toll. Hundreds of residents had moved or been displaced. The Musket Factory and Rifle Factory were in useless ruins. Even Herr's Mill had been burned because of its owner's Union sympathies. John T. Trowbridge, a New England writer who travelled through Harpers Ferry in the summer of 1865, described the sad scene:

> *But while the region presents such features of beauty and grandeur, the town is the reverse of agreeable. It is said to have been a pleasant and picturesque place formerly. The streets were well graded, and the hill-sides above were graced with terraces and trees. But war has changed all. Freshets tear down the centre of the streets, and dreary hill-sides present only ragged growths of weeds. The town itself lies half in ruins. . . .*

An even more enduring scar on the spirits of townspeople was left by political actions taken during the war. After Virginia had seceded, Union loyalists in the state's western counties chose in turn to "secede" from Virginia. Jefferson and Berkeley counties, located at the foot of the Shenandoah Valley, had closer geographical and political ties with Virginia than with the loyal western counties. But because the Union-controlled B&O Railroad travelled across these two Shenandoah Valley jurisdictions between Harpers Ferry and Cumberland, Md., political, economic and military pressure was exerted to bring both counties into the new state. On May 23, 1863, the residents of Jefferson and Berkeley counties went through the formality of a vote. But in fact, the issue was decided by federal forces occupying the region. Southern sympathizers who would not take an oath of loyalty to the Union were forced to stay home. A small turnout—only a fraction of the number who had voted in the 1860 presidential election—chose overwhelmingly to separate from Virginia. At Harpers Ferry, for instance, the count was 196 to 1. On June 20, 1863, West Virginia became the 35th state by presidential proclamation.

With the war now behind them, residents at Harpers Ferry began the long and difficult process of rebuilding their lives. Homes were repaired, shops reopened, and industry on Virginius Island made a modest comeback (see Chapter 9). But nature was not yet done with this war-ravaged town. The deforestation of the Potomac and Shenandoah river basins had proceeded steadily through most of the 1800s. Timber was cut to fuel iron furnaces and blacksmith forges, or to make room for the spread of agriculture. Damage inflicted during the war, particularly upon farms and crops in the Shenandoah Valley, left thousands of acres of barren fields. Unchecked runoff created unstable streamflow, with dangerous consequences for Harpers Ferry.

On Sept. 30 and Oct. 1, 1870, the Shenandoah River jumped its banks and swelled to a height of 29.7 feet at Harpers Ferry. Mrs. John Child, who lived with her husband on Virginius Island, described this flood in a letter to her mother and sister:

> *I snatch a few minutes to give you a few particulars of the disaster that has happened to us here. . . . We had no idea of the danger until it was too late to escape from the Island. Last Friday towards evening the water commenced rising rapidly. Before two hours every way of escape and all hope of rescue was cut off from us. So we were compelled to stay within the crumbling walls which sheltered us from the terrible water which seethed and dashed around us. There were two bridges connecting the Island with the mainland, one wooden one near our house and the railroad bridge. So violent was the water that these were torn to fragments and carried away. . . .*

Joseph Barry, in *The Strange Story of Harper's Ferry,* wrote an even more compelling account of this disaster. The plight of several residents on Overton's Island, just a few hundred yards upstream from Virginius Island, was especially tragic:

> *About this time an excited crowd had gathered at the foot of Union street, watching with intense anxiety for the fate of some families on Overton's island, directly opposite, and about sixty yards distant. Between them and the island rushed an impetuous torrent to attempt to cross which, in a boat, would be madness and the distance was too great to allow a rope of sufficient strength to be thrown to the assistance of the helpless people. The scene was truly terrible. The screams of men, women and children in imminent peril of drowning or being crushed by falling houses, and the sympathetic cries and sobs of pitying spectators were partially lost in the thunders of the furious tide. . . .*

So rapidly and with such force had the Shenandoah risen that families caught unaware were trapped in their homes. Mrs. Child and her family were lucky, for by the time the waters receded, 42 lives had been lost. In addition, the iron foundry, machine shop, saw mill and carriage manufacturing shop on Virginius Island were all levelled. John Wernwag, the son of bridgebuilder Lewis Wernwag and former partner in the firm Wernwag & Schofield, was also among the fortunate. When the machine shop, in which he was then residing, crumbled under the weight of the surging floodwaters, Wernwag made his way to the building's roof and floated safely down to Brunswick, Md.

Of the pre-war establishments on Virginius Island, only the old cotton factory—now the Child & McCreight flour mill—was left standing. Wrote Mrs. Child in her same letter:

> *The one bright spot in the picture, next to the saving of our lives, is the Mill which is comparatively uninjured. They lost several hundred barrels, some feed and one of their scales but if it were not for the destruction of the race, could go to grinding in a few days, but the race is the greatest calamity of all. The headgates are gone, walls are torn down and the race filled with sand. It will cost several thousand dollars to repair it.*

More floods followed. In 1877, floodwaters closed down the Shenandoah Canal for good, and inflicted extensive damage on the C&O Canal. On May 31 and June 1, 1889, the Potomac and Shenandoah rivers rose to their highest level yet, cresting at 34.8 feet. Joseph Barry wrote, however, that "as the currents acted as natural checks on one another, there was comparatively little damage done to property at the place, except from the filthy deposits left by the waters." Nevertheless, the Child & McCreight flour mill was finally forced to close. Only a pulp mill and, after 1890, a paper mill continued to operate at Harpers Ferry, being the last water-powered manufactories here.

The Rifle Factory ruins on Lower Hall Island shortly after the Civil War. The lower locks of the Shenandoah Canal are visible in the foreground. National Park Service

Aftermath of the Flood of 1889 on Shenandoah Street in Harpers Ferry. The Stage Coach Inn, which now serves as a National Park Service Visitor Center, is the last building on the right. National Park Service

Chapter Nine

PROSPECTS FOR RECOVERY

On July 24, 1867—two years after the Civil War had ended—Abraham H. Herr sold his Virginius Island holdings to Jonathon C. Child and John A. McCreight of Springfield, Ohio. The firm of Child & McCreight subsequently repaired and leased out the iron foundry, machine shop, saw mill and carriage manufacturing shop. The firm also proceeded to convert the old cotton factory into a flour mill. Joseph Barry, in *The Strange Story of Harper's Ferry*, described this new establishment in some detail:

> *A large force of men was kept in employment for fifteen months, preparing the building and putting up the machinery. . . . The works were of the most approved description, set in motion by four turbine wheels, the power being that of three hundred horses. There were ten run of buhrs, which turned out five hundred barrels of flour daily and, in the whole, it was said by adepts in that business, to be a marvel of ingenuity. . . .*

The "Harpers Ferry Mill Company," as the new flour mill was known, made advantageous use of mechanical contrivances first introduced by Oliver Evans in *The Young Mill-wright and Miller's Guide* in 1795 (see diagram, page 70). This apparatus included endless leather belts fitted with small buckets, chutes which directed the descent of flour by gravity, and augers which moved flour horizontally through the mill. Such devices served, in Evan's own words, "to perform every necessary movement of grain, and meal, from one part of the mill to another, or from one machine to another, through all the various operations from the time the grain is emptied from the wagoner's bag . . . until it be completely manufactured into flour. . . ." The application of water power to these various mechanisms largely replaced the need for hand labor.

The firm of Child & McCreight also rebuilt the mill's waterworks. When renovations were completed, Herr's Dam was made 400 feet long and four feet high. The headrace, more than 600 feet long, was cleaned out and thoroughly repaired. Altogether, these new waterworks provided the flour mill with a head and fall of 14 feet.

The "Harpers Ferry Mill Company," as previously mentioned, was slightly damaged during the Flood of 1870; it was the mill's headrace, headgates, and dam which suffered the worst effects of the raging river. Repairs were promptly made, however, and the flour mill was put back in operation in a few months. The Flood of 1889 also inflicted extensive damage on the mill's dam and headrace. But this time, repairs were not attempted. After more than 20 years of operation, the flour mill was finally abandoned.

Following the Civil War, the government never seriously considered rebuilding the Harpers Ferry Armory. This arms manufactory had always consumed considerable amounts of capital, and had never attained the economy of production reached at the Springfield Armory. Consequently, an act of Congress, passed on Dec. 15, 1868, provided for the sale of all government property at Harpers Ferry. About a year later, on Nov. 30 and Dec. 1, 1869, this property was disposed of at public auction. Most armorers' dwelling houses and other small lots were acquired by local residents. The ferry rights to the Shenandoah River, owned by the government since 1818, were also sold at this time. The armory grounds along the Potomac River and the rifle factory site above Virginius Island, the two largest government tracts, were purchased by Capt. F.C. Adams of Washington, D.C.

Townspeople were indeed hopeful that Adams would erect new industry and create new jobs at Harpers Ferry. But in fact, Adams only intended to challenge the B&O Railroad's right-of-way through the old armory grounds on some legal technicality. In this manner, he hoped to obtain an attractive financial settlement. But the effort failed, and Adams consequently withheld payment for the property. After considerable litigation, the armory grounds and rifle factory site reverted back to the government in 1876.

A second public auction was subsequently authorized by Congress and held at Harpers Ferry on Oct. 22, 1884. At this time, Thomas H. Savery of Wilmington, Del., purchased the armory grounds, rifle factory site, riparian land, and flowage rights to the Potomac and Shenandoah rivers—about 200 acres in all—for $200,000. Two years later, the Shenandoah Pulp Company was organized "for the purpose of manufacturing wood pulp and for the transaction of business incident thereto." Savery was the company's first president, and John F. Quigley was its general manager. On Wednesday, April 20, 1887, Quigley began erecting the Shenandoah Pulp Company mill on the site of the lower locks of the old Shenandoah Canal.

It is quite possible that Quigley read the government's *Reports on the Water-Power of the United States,* issued in 1885. In this document, the availability of water power at Virginius Island was

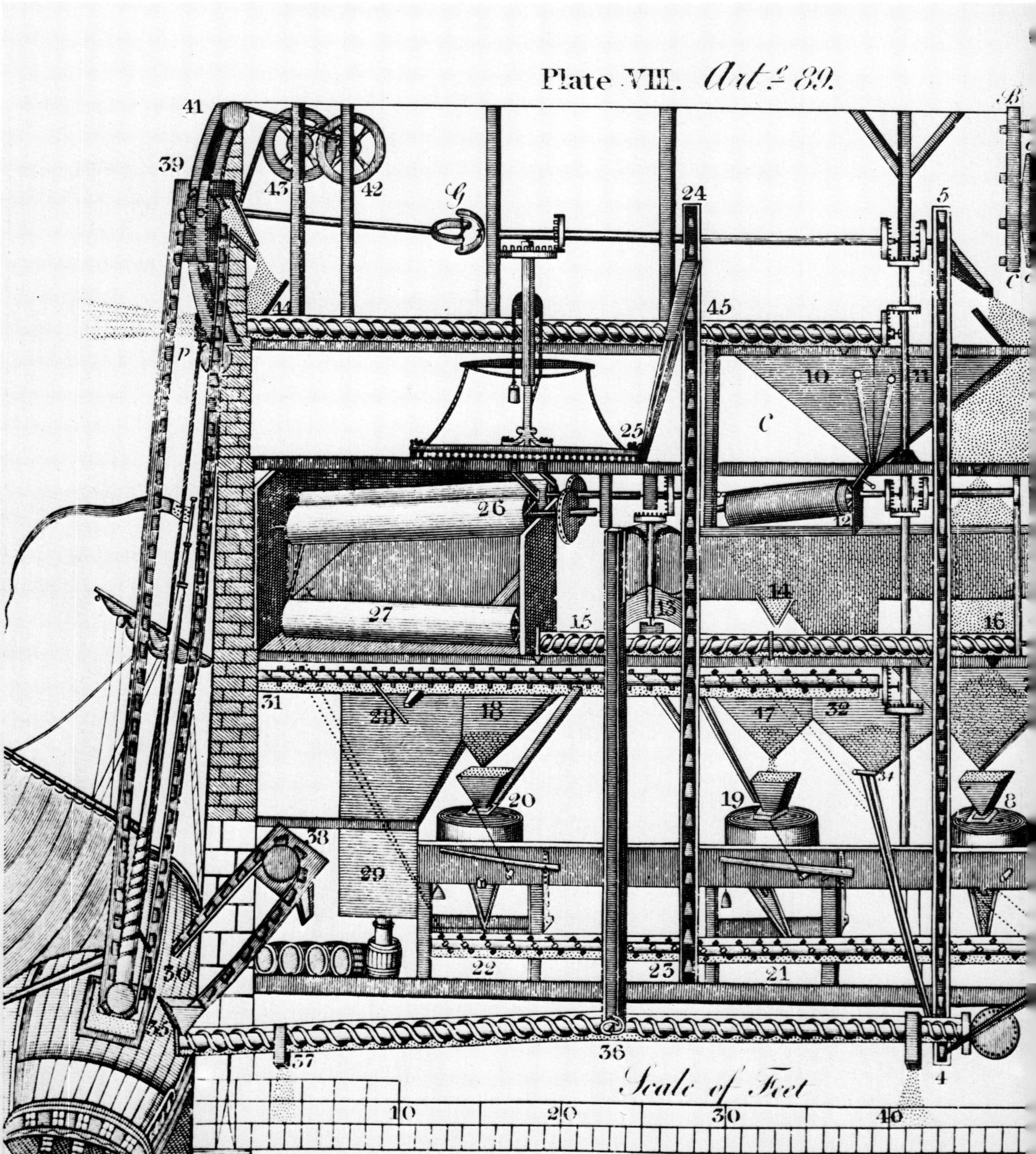

Illustration of a flour mill equipped with Oliver Evans' various mechanical contrivances, from The Young Mill-wright and Miller's Guide *(8th edition, published in 1834). Elevators, augers, and descenders conveyed grain from process to process without the need for hand labor.* National Park Service

specifically addressed:

> *By building a dam at the head of the island and turning the water into the old canal, this fall could be used without difficulty, and with considerable building-room on the island; and a much greater fall than 14 feet could be used, if necessary, depending on the height of the dam.*

By the middle of July 1887, Quigley had rebuilt the Shenandoah Dam about a mile upstream from the unfinished pulp mill. The new dam was of typical construction: timber piers, five to six feet high, were bolted together with iron spikes and filled with stone rip-rap. New headgates were also erected, through which water was conveyed into the old Shenandoah Canal and down to the pulp mill. By this arrangement, a head and fall of about 25 feet was obtained—almost twice that of the nearby Child & McCreight flour mill. A large impoundment behind the mill, called, appropriately, Lake Quigley, temporarily stored the water until it was needed to generate power.

Work on the pulp mill proceeded according to plan until July 19, 1887. On that date, the "Harpers Ferry Mill Company" filed suit against "Thos. H. Savery and others," charging that the diversion of water by the new Shenandoah Dam would impinge upon the flowage rights obtained by Jonathon Child and John McCreight from Abraham H. Herr 20 years earlier. Because water supplied to the pulp mill would not be returned to the Shenandoah River until it had passed the Child & McCreight flour mill, these owners argued that the effectiveness of their four turbines would be significantly impaired. The courts temporarily halted construction of the pulp mill and examined the deeds of both companies. By the end of the summer, however, the courts ruled in Savery's favor, upholding the flowage rights accorded him by the government.

In 1888, 10 turbines were installed in the pulp mill's five flumes. Three pair were New American Water Wheels on horizontal shafts (see diagram, page 76), one pair were Improved Success Water Wheels on horizontal shafts, and the last pair were New American Water Wheels on vertical shafts. Together, these turbines delivered 1,600 horsepower to the mill's pulp-making machinery.

During most of the 1800s, the basic raw material for paper was cotton and linen rags. Rag paper was durable and attractive, but as early as the 1850s, the supply of rags could not keep pace with the demand for paper. The idea of using wood as a raw material for paper had been considered early in the 1700s. But it wasn't until about 1858, when German inventor Henry Voelter received a patent from the United States for a new wood-pulp grinder, that a mechanical process for reducing wood to fiber was adopted in this country.

Ruins of the brick-lined arches that formerly admitted water from the Shenandoah River to the mills and factories on Virginius Island.

The Shenandoah Pulp Company used this mechanical process to manufacture ground wood pulp—the first stage in the paper-making process. Logs were lifted into the mill from Lake Quigley by a log hoist, stripped of their bark, and cut into three-foot lengths. These short logs were then loaded into a grinder in the wood grinder room. Each wood-pulp grinder had small doors on its front side, through which the logs were introduced to a cylindrical grinding stone (see diagram, page 76). The stone was fitted on a horizontal spindle, which in turn was coupled to the horizontal shaft of a pair of turbines. While the stone turned, mechanical pushers, operated by screw feeds or hydraulic feeds, caused the logs in the grinding pockets to advance constantly and regularly against the stone. In this manner, the logs were ground into pulp fiber.

From the grinder, pulp fibers were conveyed through pipes to a wet machine. Here, in an oak screen cradle, the pulp was mechanically mixed and suspended in water. A system of shower pipes also washed the pulp fibers from unground slivers of wood. The pulp was then dropped onto a "Gould screen"—an endless moving wire cloth comprised of fine meshes. Here, the pulp fibers were collected while water was allowed to pass through. From the wire cloth, the pulp passed through pressing rolls and drying cylinders, the latter heated by steam. In this manner, the pulp was rolled into strips. These pulp strips were then loaded into box cars for shipment. In 1927, the capacity of the Shenandoah Pulp Company mill was 15 tons of ground wood pulp daily.

Workers operating the wood-pulp grinders in Savery's mill earned just $1 for a 10-hour day in 1891. Those men attending to the rolling machines were paid $1.25 for a 12-hour day. Twenty-five years later, in 1915, wages for a 12-hour day had risen by only 25 cents—to $1.50.

In 1890, Savery completed the Harpers Ferry Paper Company mill along the Potomac River on the site of the armory's old rolling mill. This factory, like the pulp mill, was also fitted with 10 turbines—four pair of New American Water Wheels on horizontal shafts and one pair of Improved Success Water Wheels on vertical shafts. Two more turbines on vertical shafts were added later—one in 1910, and one in 1923. These several turbines powered machinery much like that contained in the Shenandoah Pulp Company mill.

Savery used the old Armory Dam and Armory Canal to convey water to his paper mill. The total head and fall here was about 24 feet. But again Savery encountered difficulties over the flowage rights he had presumably obtained from the government. Contradictory deeds in fact granted conflicting rights to Savery and the C&O Canal for water from the Potomac River. Extensive correspondence was generated between both parties, but in this instance the Canal Company did not press the matter.

On July 28, 1891, Savery sold John Brown's Fort, which stood on property he had obtained from the government, to a world exposition in Chicago. By the end of the year, the building was dismantled and shipped westward. The following year, the B&O Railroad purchased a right-of-way from Savery for a new track arrangement at Harpers Ferry. Between 1892-94, the railroad company cut a tunnel over 800 feet long through a spur of Maryland Heights, erected a new bridge across the Potomac River, and covered the armory grounds with a 20-foot embankment. The new bridge and track arrangement were put into service on April 12, 1894—replacing the B&O's original mainline which had run along the Potomac riverwall. Virtually all signs of the old Musket Factory were now gone.

Savery had erected his two mills at Harpers Ferry during a period of economic growth and opportunity. The general adoption of Voelter's wood-pulp grinder and the introduction of other new machines and methods for manufacturing wood pulp stimulated the rapid expansion of America's entire pulp and paper industry, especially between 1870-1885. By 1890, paper mills in this country were annually consuming some 225,000 tons of mechanical pulp. Also in about 1890, a chemical process for pulping wood, using sulphur in lead digesters to separate the wood fibers, was perfected.

Characteristically, rapid expansion resulted in over-production and falling prices as new mills battled for a share of the market. In 1865, for example, print paper sold for 24-1/2 cents per pound; by the

Herr's Dam and the three water gates still conveyed water to the Child & McCreight flour mill when this photo was taken in about 1882. The ruins of Herr's Mill, burned by Confederate raiders in 1861, are visible in front of the water gates. National Park Service

This circa 1888 photograph shows the new Shenandoah Pulp Company mill and Lake Quigley behind it. The "Wood Grinder Room" occupied the lower level of the mill directly over the five tailrace flumes.

National Park Service

Ruins of the pulp mill, after the Blizzard of 1983.

early 1890s, the price had fallen to 3 cents per pound. Only the largest pulp and paper producers, due to their volume of output, were able to make any profit.

The Shenandoah Pulp Company and Harpers Ferry Paper Company, built during this nationwide fit of expansion, were not among these profitable manufactories. In 1904, for instance, the pulp mill tried to make up for mounting losses by leasing the use of one of its turbines to the Harpers Ferry Electric Light and Power Company for $600 a year. Still, the pulp mill's debts mounted—from 1924 to 1927, the Shenandoah Pulp Company lost over $72,000.

Meanwhile, the Harpers Ferry Paper Company mill, which also struggled with growing losses, was totally destroyed by fire on Jan. 15, 1925. Although the pulp mill endured for another 10 years, it did little more than keep a handful of townspeople employed. Finally, without fanfare, the Shenandoah Pulp Company mill closed down in August 1935. Altogether, the pulp mill had operated for 45 continuous years—longer than any other establishment at Harpers Ferry except the armory. The all-time record flood of March 18 and 19, 1936, during which the rivers rose 36.5 feet, swept the abandoned pulp mill away, leaving only its massive stone foundations.

The Shenandoah Pulp Company was the last water-powered mill to operate at Harpers Ferry. The armory had come and gone; the Flood of 1870 had swept Virginius Island clean of most of its industry; and the Flood of 1924 toppled the last remaining walls of the old cotton factory. Only a small hydro-electric power plant, built on the foundations of the paper mill by the Harpers Ferry Electric Light and Power Company in 1927, continued to use water power. Today, this facility still uses two turbines to generate up to 840 kilowatts of power for local households.

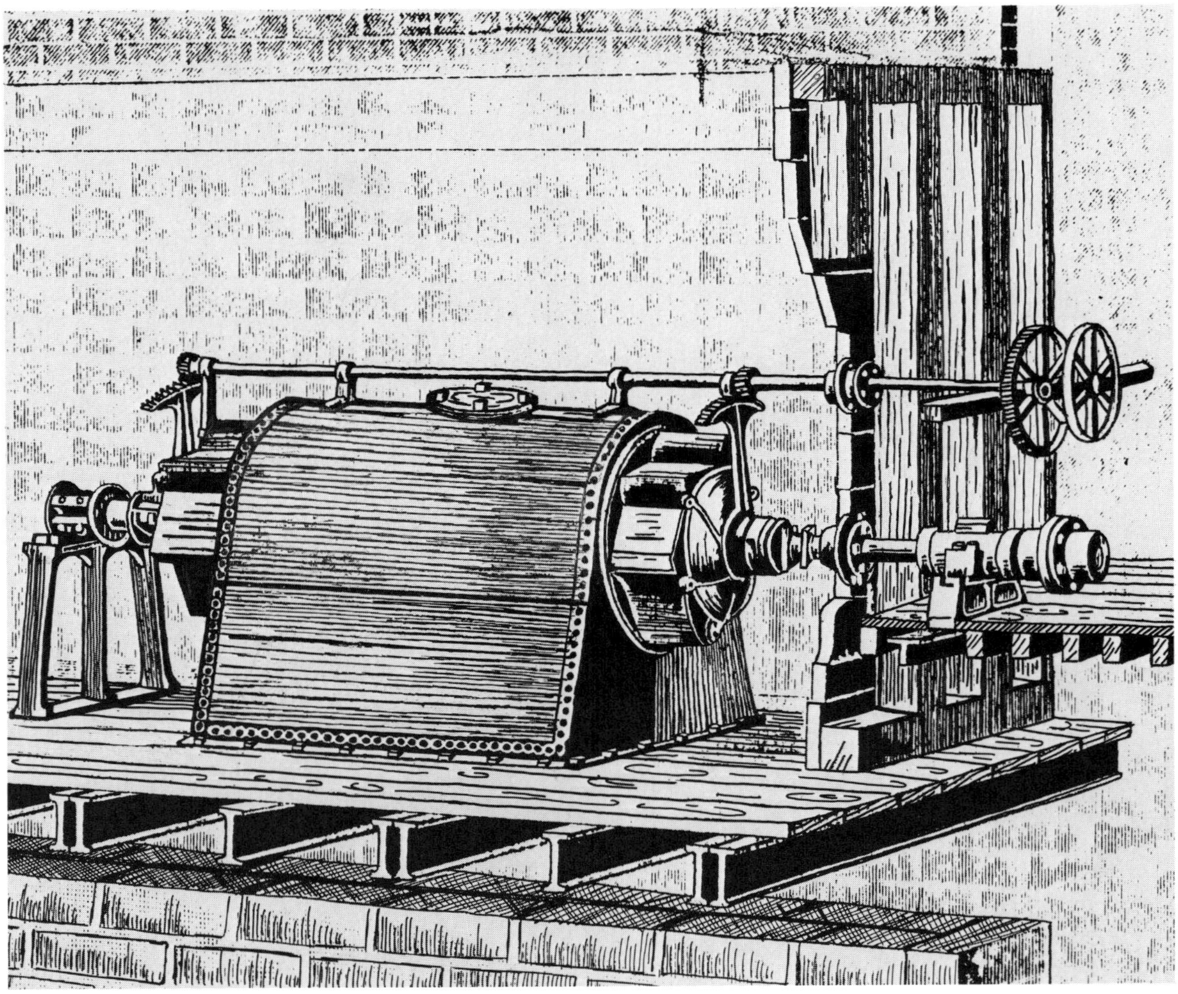

Top: A pair of New American turbines on a horizontal shaft. According to an 1892 catalog of The Dayton Globe Iron Works, this illustration "was taken from an outfit we furnished the Shenandoah Pulp Co., of Harper's Ferry, West Virginia, to work under a 24-foot head."

Drawing reproduced by Marc Melcher

Right: Front view of an improved Voelter wood-pulp grinder, patented in 1866. This machine had five grinding pockets and automatic screw feeds which advanced the logs against the grinding stone. The horizontal spindle could be coupled directly to the horizontal shaft of a pair of turbines.

Protruding end of a horizontal shaft from a pair of New American turbines in a flume at the old paper mill. This shaft was coupled directly to a wood-pulp grinder in the mill's wood grinder room, located directly in front of the flume. The handwheel above the shaft opened and closed the gates around each turbine.

Circa 1900 photograph of the Harpers Ferry Paper Company mill. Logs, waiting to be hoisted into the mill and ground into pulp, are visible behind the mill on the Armory Canal. National Park Service

Chapter Ten

DECLINE OF WATER POWER

It is hard to find a single reason why industry failed at Harpers Ferry. From the days of great promise when Robert Harper first settled at The Hole, to the unceremonious closing of the Shenandoah Pulp Company in 1935, a combination of factors contributed to the demise of water-powered industry here.

Unreliable streamflow and wide fluctuations in water levels, for instance, became more and more alarming as the 19th century progressed. Prolonged periods of low flow sometimes impaired the function of heavy machinery, and actually forced a few mills to close down for days or even weeks at a time. The Child & McCreight flour mill, for example, did not generally operate during the winter months, when low water was relatively common and ice often blocked its headgates. The records of the Shenandoah Pulp Company also show that several working days were lost every year due to an inadequate supply of water for power. Because the development of America's factory system depended so much on increased mechanization, the unreliability of the Potomac and Shenandoah rivers became a substantial obstacle to the productivity of individual establishments.

Periodic high water, on the other hand, inflicted extensive damage on millraces and milldams. The government, for example, was forced to expend large sums of money for periodic repairs to the Armory Dam and Armory Canal. As mentioned previously, some $68,000 was spent on repairs and improvements to the Armory Canal alone between 1830-1860. The firm of Child & McCreight likewise spent a considerable sum of money rebuilding Herr's Dam and the headgates for their flour mill, which were completely ruined during the Flood of 1870. The Shenandoah Pulp Company was also forced to make frequent repairs on the Shenandoah Dam. On July 6, 1922, for instance, the company's general manager reported that new logs with tie logs were attached to the top of the dam, and that the structure was made tight with stone rip-rap and gravel. The manager then summed up the situation: "This will hold water for a time, until we decide to rebuild the dam."

Major floods inflicted extensive damage not only on milldams, headgates, and millraces, but destroyed several manufactories as well. The most prominent victims of floodwaters were the cotton mill in the Flood of 1852; the iron foundry, machine shop, saw mill, and carriage manufacturing shop in the Flood of 1870; and the Child & McCreight flour mill, which was forced to close after the Flood of 1889.

Fire also took its toll on industry at Harpers Ferry. The original "Island Mill" on Virginius Island burned to the ground in 1839. The Harpers Ferry Paper Company mill along the Potomac River was totally destroyed by fire in 1925.

The Civil War certainly inflicted the most costly destruction, in terms both of physical damage and jobs lost. The expansive government Musket Factory and Rifle Factory, which together employed about 400 men in 1860, were burned during the very first year of the war. Herr's Mill was also put to the torch by Confederate raiders in 1861.

Nationwide economic factors played a prominent role in the failure of two particular manufactories. The financial troubles of the Harpers Ferry and Shenandoah Manufacturing Company, which had over-extended itself during boom years in the textile industry, forced a court-ordered public sale of the cotton factory in 1853. Similarly, over-expansion and over-production in the pulp and paper industry during the late 1800s saddled Thomas Savery's two Harpers Ferry mills with mounting debts. Ultimately, these financial hardships forced the Shenandoah Pulp Company to close in 1935. Had the Harpers Ferry Paper Company not burned to the ground 10 years earlier, it too would inevitably have met the same fate.

Clearly, there were several ingredients that combined to bring ruin to water-powered industry at Harpers Ferry. Each ingredient, taken separately, had a substantial negative impact on one establishment or another. Together, these several ingredients proved insurmountable to the health of local industry as a whole. But another factor, applied in a much wider context, had for several decades been eating away at the very foundations of water-powered industry here and elsewhere.

In 1769, Scottish engineer James Watt patented an efficient low-pressure steam engine—a considerable improvement over an earlier model developed in 1705 by Thomas Newcomen in England. Watt, in partnership with Mathew Boulton, soon began the commercial manufacture of steam engines. By 1803, both men were exporting their Boulton-Watt engines to the United States. Also in about 1803, Oliver Evans developed a simple, compact, high-pressure steam engine for either stationary use or steamboat use. Four years later, in 1807, Evans installed one of his engines in a mill in Lexington, Ky.—the first steam-powered mill in

America.

The advantages of steam power over water power were readily apparent. Steam engines could produce power reliably and effectively without regard to the fluctuations in streamflow which so impaired water-powered mills. Steam-powered mills, because they did not rely on adjacent rivers for power, could also be erected virtually anywhere—near large urban markets or raw materials, for instance.

Still, steam power had its drawbacks, and by 1840 it accounted for no more than 10 to 15 percent of all stationary power produced in this country. Steam engines, for example, were much more expensive to obtain in the early 1800s than milldams, millraces, and conventional water wheels were to build. In addition to the high purchase cost, the difficulty and expense of transporting heavy steam engines prior to the extension of railroad service was extremely prohibitive. Fuel for these engines—first charcoal and subsequently coal—was relatively expensive compared to free-flowing water. Millwrights and mill owners were also unfamiliar with the operation, maintenance, and repair of steam engines during these early years; by contrast, the craft skills, necessary tools, and appropriate building materials to erect milldams and conventional water wheels were available in even the most remote communities.

As steam engines became more efficient, less costly, and more accessible to manufactories served by railroad, the economics of steam power became increasingly attractive. By the mid-1850s, steam power cost about three times as much as water power; a decade later, this margin had been narrowed to about two to one. Indeed, not only was the cost of steam power falling—the cost of water power was rising.

Higher land costs and property taxes became an increasing burden on mill owners who needed considerable land for long headraces and large milldams. The maintenance and repair of these same waterworks, which were regularly damaged by high water or clogged with silt and other debris, also required constant attention and expense. By contrast, steam engines, although they still required large initial outlays of capital, proved themselves durable and relatively maintenance free.

Following the Civil War, the gap between steam power and water power was gradually closed. In 1870, the federal decennial census found that steam horsepower actually led water horsepower

The National Park Service has stabilized and restored many of the old mercantile buildings in the Lower Town of Harpers Ferry. This particular building, erected in 1844, was occupied by a butcher and grocery store at the time of John Brown's raid.

St. Peter's Roman Catholic Church at Harpers Ferry. The original church was built in 1830-33 on land donated by the Wager family. The present building replaced the original structure in 1896. The Harper House is visible on the right.

in manufacturing, by a ratio of about 52 to 48. Ten years later, in 1880, steam finally passed water as a power source both in power produced and in number of units.

The declining economy of water power, then, proved in the long run to be the root cause for the failure of industry at Harpers Ferry. The owners of water-powered mills found themselves saddled with a less reliable, increasingly expensive form of energy by the last decades of the 19th century. Under these circumstances, they simply could not effectively compete with more efficient, reliable, and productive steam-powered factories. While war, fire, and flood had indeed exacted a heavy physical and economic toll on individual establishments, the gradual decline of water power as an economical means of producing goods had already undermined the financial foundations of these same manufactories. The high cost of producing arms at the Harpers Ferry Armory, for instance, and the unsatisfactory economy of production obtained here can be attributed in part to the never-ending expense of maintaining the Armory Dam and Armory Canal, and to inadequate streamflow, which sometimes impaired the function of important machinery. These considerations were a prominent factor in the government's decision not to rebuild the armory workshops after the Civil War.

Similarly, there was surprisingly little effort by private investors to resurrect industry at Harpers Ferry on any scale even approaching that which had preceded the Civil War. It was an interesting coincidence that in 1870, a disastrous flood swept away most of the remaining water-powered mills on Virginius Island, and steam horsepower first passed water horsepower in manufacturing nationwide. Just as the growth of industry at Harpers Ferry had reflected the development of America's expanding factory system, so, too, the demise of water-powered mills here—hurried along by war, fire, and flood—mirrored the economic decline of water power across the country.

The failure of industry, and of man's earnest efforts to control his environment at Harpers Ferry, provide us with a valuable lesson. Despite decades of unchecked development, extensive deforestation, and the construction of substantial waterworks to rechannel the rivers, nature has endured. Compare Harpers Ferry today with old photographs of the Musket Factory, Rifle Factory, and Virginius Island. No longer do mills, factories, and dwelling houses dominate the scene. Instead, nature has reclaimed the shorelines and hillsides with amazing thoroughness. The Potomac and Shenandoah rivers again run freely down their rocky beds. Woodland communities boast a prolific variety of birds, mammals, and flora. Only scattered ruins—an old factory foundation here, an abandoned millrace there—remind visitors today that industry once proliferated at Harpers Ferry. John Brown's Fort also provides a link to the past. But the building has been dismantled and moved so often—four times, since Thomas Savery first sold the landmark in 1891—that it barely resembles the armory's original fire enginehouse and watchman's office in form or character.

With enthusiasm, we can once more look out from Jefferson Rock above the Lower Town of Harpers Ferry and remember the words of Thomas Jefferson, written 200 years ago: "The passage of the Patowmac through the Blue Ridge is perhaps one of the most stupendous scenes in Nature."

Factories were not the only structures to suffer damage at Harpers Ferry. This particular bridge at the mouth of the Shenandoah River, erected in 1882, was swept away once by the Flood of 1889, rebuilt, and later destroyed for good by the record Flood of 1936.

RESEARCH SOURCES

Bagnall, William R., *The Textile Industries of the United States.* 1893. Reprinted in New York by Augustus M. Kelley, 1971.

Barry, Joseph, *The Strange Story of Harper's Ferry.* Shepherdstown, W.Va.: The Shepherdstown Register, 1903.

Becker, Carl M., "James Leffel: Double Turbine Water Wheel Inventor," *Ohio History,* Vol. 75, No. 4, Autumn 1966, 200-211.

Brooks, Maurice, *The Appalachians.* Grantsville, W.Va.: Seneca Books, 1965.

Bushong, Millard Kessler, *Historic Jefferson County.* Boyce, Va.: Carr Publishing Company, 1972.

Catalogue of the New American Turbine, Manufactured by The Dayton Globe Iron Works Co., Successors to Stout, Mills & Temple. Dayton, Ohio, 1892.

Clark, Victor S., *History of Manufactures in the United States, Vol. 1, 1607-1860.* New York: Peter Smith, 1949.

_______________, *History of Manufactures in the United States, Vol. 2, 1860-1893.* New York: Peter Smith, 1949.

Cloos, Ernst, *Structural Geology of South Mountain and Appalachians in Maryland.* Baltimore: The Johns Hopkins Press, 1958.

_______________, *The Geology of the South Mountain Anticlinorium, Maryland.* Baltimore: The Johns Hopkins Press, 1950.

Davis, Charles Thomas, *The Manufacture of Paper.* Philadelphia: Henry Carey Baird & Company, 1886.

Davis, Julia, *The Shenandoah.* New York: Farrar & Rinehart, 1945.

Douglas, Henry Kyd, *I Rode with Stonewall.* Chapel Hill: University of North Carolina Press, 1940.

Ernst, Ora Ann, "Mill Spanning Two Centuries Still Stands at Big Spring," *The Daily Mail,* Hagerstown, Md., Saturday, July 3, 1976.

Evans, Oliver, *The Young Mill-wright and Miller's Guide,* 13th Edition. Philadelphia: Lea & Blanchard, 1850. Reprinted in New York by Arno Press, 1972.

Everhart, William C., *A History of Harpers Ferry.* Harpers Ferry: National Park Service, 1952.

Faulkner, Harold Underwood, *American Economic History.* New York: Harper & Row, 1960.

Fox, William, Bill Brooks and Janice Tyrwhitt, *The Mill.* Boston: New York Graphic Society, 1976.

Freeman, Douglas Southall, *George Washington, Vol. 1.* New York: Charles Scribner's Sons, 1948.

_______________, *George Washington, Vol. 6.* New York: Charles Scribner's Sons, 1954.

Gutheim, Frederick, *The Potomac.* New York: Grosset & Dunlop, 1968.

Hahn, Thomas F., *George Washington's Canal at Great Falls, Virginia.* Shepherdstown, W.Va: American Canal & Transportation Center, 1976.

_______________, *Towpath Guide to the C&O Canal.* Shepherdstown, W.Va.: American Canal & Transportation Center, 1982.

Hannah, David, *Archeological Excavations on Virginius Island, 1966-68.* Harpers Ferry: Job Corps Civilian Conservation Center, 1969.

Hindle, Brooke, editor, *America's Wooden Age: Aspects of its Early Technology.* Tarrytown, N.Y.: Sleepy Hollow Restorations, 1976.

Hungerford, Edward, *The Story of the Baltimore & Ohio Railroad, 1827-1927.* Vols. 1-2. New York: G.P. Putnam's Sons, 1928.

Illustrated and Descriptive Pamphlet and Price List of The Improved Success Water Wheel, Manufactured by S. Morgan Smith. York, Pa., c. 1889.

Interstate Commission on the Potomac River Basin, *Water in the Great Valley: People, Problems, Decisions.* Rockville, Md.: ICPRB, 1980.

Jefferson County Historical Society, *Magazine of the Jefferson County Historical Society,* Vol. VIII, December 1942, 24.

John Brown's Raid. Washington, D.C.: National Park Service, 1973.

Kuethe, J. Lewis, "A List of Maryland Mills, Taverns, Forges, and Furnaces in 1795," *Maryland Historical Magazine,* Vol. 31, 1936, 155-169.

Kytle, Elizabeth, *Home on the Canal.* Cabin John, Md.: Seven Locks Press, 1983.

Land Records of Jefferson County, Virginia.

McGrain, John W., "Mills of Washington County, Maryland." 1967.

Mead, Daniel W., *Water Power Engineering.* New York: McGraw-Hill Book Company, 1915.

Murfin, James V., *The Gleam of Bayonets.* New York: Bonanza Books, 1965.

The National Register of Historic Places, 1972. Washington, D.C.: U.S. Government Printing Office, 1972.

Putnam, William C., *Geology.* New York: Oxford University Press, 1964.

Reports on the Water-Power of the United States, Census Office, Department of the Interior, Part 1. Washington, D.C.: Government Printing Office, 1885.

Reynolds, John, *Windmills & Watermills.* New York: Praeger Publishers, 1970.

Rothfuss, Joel F., *Geological History of the Harpers Ferry Area.* Harpers Ferry: National Park Service, 1970.

Sanderlin, Walter S., *The Great National Project, A History of the Chesapeake & Ohio Canal.* Baltimore: The Johns Hopkins Press, 1946.

Sarles, Frank B., Jr., *Economic and Social History of Virginius Island.* Washington, D.C.: National Park Service, 1969.

Savery Papers. Harpers Ferry National Historical Park.

Scharf, J. Thomas, *History of Western Maryland, Vol. 2.* Baltimore: Regional Publishing Company, 1968.

Sloane, Eric, *Our Vanishing Landscape.* New York: Funk & Wagnalls, 1955.

Smith, Merritt Roe, *Harpers Ferry Armory and the New Technology.* Ithaca, N.Y.: Cornell University Press, 1977.

Smith, Philip R., Jr., *History of Loudoun Heights, 1813-1880.* Harpers Ferry: National Park Service, 1959.

Snell, Charles W., *A Compendium of the Commercial and Industrial Advertisements of the Business and Manufacturing Establishments of Harpers Ferry and the Island of Virginius, 1824-1861, Virginia.* Denver: National Park Service, 1973.

_______________, *A Short History of the Island of Virginius, 1816-1870.* Harpers Ferry: National Park Service, 1959.

_______________, *The Business Enterprises and Commercial Development of Harpers Ferry's Lower Town Area, 1803 to 1861.* Harpers Ferry: National Park Service, 1973.

_______________, and Philip R. Smith, Jr., *The Musket Factory Buildings and Grounds, Harpers Ferry Armory, 1859-1860.* Harpers Ferry: National Park Service, 1959.

Stinson, Dwight E., *The First Railroad Bridge at Harpers Ferry.* Harpers Ferry: National Park Service, 1970.

Sullivan, Arthur L., *Harpers Ferry in the Civil War, 1862.* Harpers Ferry: National Park Service, 1961.
Summers, Festus P., *The Baltimore and Ohio in the Civil War.* New York: G.P. Putnam's Sons, 1939.
Thompson, Michael D., *The Iron Industry in Western Maryland.* 1976.
United States Patent Office Records.
Vokes, Harold E., *Geography and Geology of Maryland.* Baltimore: Maryland Geological Survey, 1961.
Water Resources Data for Maryland & Delaware, Water Year 1979. Towson, Md.: U.S. Geological Survey, Water Resources Division, 1980.
Water Resources Data for West Virginia, Water Year 1979. Charleston, W.Va.: U.S. Geological Survey, Water Resources Division, 1980.
Wellman, Manly Wade, *Harpers Ferry: Prize of War.* Charlotte, N.C.: McNally of Charlotte, 1960.

INDEX